Lifetime Data: Statistical Models and Methods

SERIES ON QUALITY, RELIABILITY AND ENGINEERING STATISTICS

Series Editors: *M. Xie (National University of Singapore)*
T. Bendell (Nottingham Polytechnic)
A. P. Basu (University of Missouri)

Published

Vol. 3: Contributions to Hardware and Software Reliability
P. K. Kapur, R. B. Garg and S. Kumar

Vol. 4: Frontiers in Reliability
A. P. Basu, S. K. Basu and S. Mukhopadhyay

Vol. 5: System and Bayesian Reliability
Y. Hayakawa, T. Irony and M. Xie

Vol. 6: Multi-State System Reliability
Assessment, Optimization and Applications
A. Lisnianski and G. Levitin

Vol. 7: Mathematical and Statistical Methods in Reliability
B. H. Lindqvist and K. A. Doksum

Vol. 8: Response Modeling Methodology: Empirical Modeling for Engineering and Science
H. Shore

Vol. 9: Reliability Modeling, Analysis and Optimization
H. Pham

Vol. 10: Modern Statistical and Mathematical Methods in Reliability
A. Wilson, S. Keller-McNulty, Y. Armijo and N. Limnios

Vol. 11: Life-Time Data: Statistical Models and Methods
J. V. Deshpande and S. G. Purohit

Vol. 12: Encyclopedia and Handbook of Process Capability Indices: A Comprehensive Exposition of Quality Control Measures
W. L. Pearn and S. Kotz

Vol. 13: An Introduction to the Basics of Reliability and Risk Analysis
E. Zio

Vol. 14: Computational Methods for Reliability and Risk Analysis
E. Zio

Vol. 15: Basics of Reliability and Risk Analysis: Worked Out Problems and Solutions
E. Zio, P. Baraldi and F. Cadini

Vol. 16: Lifetime Data: Statistical Models and Methods (2nd Edition)
J. V. Deshpande and S. G. Purohit

The complete list of the published volumes in the series can be found at
http://www.worldscientific.com/series/sqres

Series on Quality, Reliability and Engineering Statistics Vol. 16

Lifetime Data: Statistical Models and Methods

Second Edition

Jayant V. Deshpande & Sudha G. Purohit
University of Pune, India

World Scientific

NEW JERSEY • LONDON • SINGAPORE • BEIJING • SHANGHAI • HONG KONG • TAIPEI • CHENNAI • TOKYO

Published by

World Scientific Publishing Co. Pte. Ltd.

5 Toh Tuck Link, Singapore 596224

USA office: 27 Warren Street, Suite 401-402, Hackensack, NJ 07601

UK office: 57 Shelton Street, Covent Garden, London WC2H 9HE

Library of Congress Cataloging-in-Publication Data
Names: Deshpande, J. V., author. | Purohit, Sudha G., author.
Title: Lifetime data : statistical models and methods /
by Jayant V. Deshpande, Sudha G. Purohit.
Description: 2nd edition. | New Jersey : World Scientific, 2016. |
Series: Series on quality, reliability, and engineering statistics ; vol. 16 |
Includes bibliographical references and index.
Identifiers: LCCN 2015031961 | ISBN 9789814730662 (hardcover : alk. paper)
Subjects: LCSH: Failure time data analysis. | Survival analysis (Biometry) |
Reliability (Engineering)--Statistical methods.
Classification: LCC QA276 .D446 2016 | DDC 519.5/46--dc23
LC record available at http://lccn.loc.gov/2015031961

British Library Cataloguing-in-Publication Data
A catalogue record for this book is available from the British Library.

Printed in Singapore

Contents

Preface to the Second Edition

The first edition of the book appeared in 2005. The copies are not easily available now. The authors and the publisher are happy to bring out a revised edition.

We have taken this opportunity to correct several typographical errors and also expand the material in various chapters. In particular, Chapter 8, which discussed only the proportional hazards model, now also includes the 'Frailty Model' as well as the 'Accelerated Life Time Model'. It has now been renamed.

This book has been used as a textbook at the Master's level at many institutions. The teachers have communicated typographical errors which they encountered. We are grateful to all of them and particularly to Prof. Isha Dewan of the Indian Statistical Institute for this curtsey.

Preface to the First Edition

The last 50 years have seen a surge in the development of statistical models and methodology for data consisting of lifetimes. This book presents a selection from this area in a coherent form suitable for teaching postgraduate students. In particular, the background and needs of students in India have been kept in mind.

The students are expected to have adequate mastery over calculus and introductory probability theory, including the classical laws of large numbers and central limit theorems. They are also expected to have undergone a basic course in statistical inference. Certain specialized concepts and results such as U-statistics limit theorems are explained in this book itself. Further concepts and results, e.g., weak convergence of processes and martingale central limit theorem, are alluded to and exploited at a few places, but are not considered in depth.

We illustrate the use of many of these methods through the commands of software R. The choice of R was made because it is in public domain and also because the successive commands bring out the stages in the statistical computations. It is hoped that users of statistics will be able to choose methods appropriate for their needs, based on the discussions in this book, and will be able to apply them to real problems and data with the help of the R-commands.

We have taught courses based on this material at the University of Pune and elsewhere. It is our experience that most of this material can be taught in a one-semester course (about 45–50 one-hour lectures over 15/16 weeks). Lecture notes prepared by the authors for this course have been in circulation at Pune and elsewhere for several years. Inputs from colleagues and successive batches of students have been useful in finalizing this book. We are grateful to all of them. We also record our appreciation of the support received from our families, friends and all the members of the Department of Statistics, University of Pune.

Chapter 1

Introduction

It is universally recognized that lifetimes of individuals, components, systems, etc. are unpredictable and random, and hence amenable only to probabilistic and statistical laws. The development of models and methods to deal with such random variables took place in the second half of the 20-th century, although certain explicit and implicit results are from earlier times as well. The development proceeded in two main intermingling streams. The reliability theory stream is concerned with models for lifetimes of components and systems, in the engineering and industrial fields. The survival analysis stream mainly drew inspiration from medical and similar biological phenomena. In this book we bring the two streams together. Our aim is to emphasise the basic unity of the subject and yet to develop it in its diversity.

In all the diverse application, the random variable of interest is the time upto the occurrence of the specified event often called "death", "failure", "break down" etc. It is called the lifetime of the concerned unit. However, there are situations where the technical term "time" does not represent time in the literal sense. For example, it could be the number of operations a component performs before it breaks down. It could even be the amount that a health insurance company pays in a particular case.

Examples of failure or lifetime situations:

(1) A mechanical engineer conducts a fatigue test to determine the expected life of rods made of steel by subjecting n specimens to an axial load that causes a specified stress. The number of cycles are recorded at the time of failure of every specimen.

(2) A manufacturer of end mill cutters introduces a new ceramic cutter material. In order to estimate the expected life of a cutter, the manu-

facturer places n units under test and monitors the tool wear. A failure of the cutter occurs when the wear-out exceeds a predetermined value. Because of the budgeting constraints, the manufacturer runs the test for a month.

(3) A 72-hr. test was carried out on 25 gizmos, resulting in r_1 failure times (in hrs.). Of the remaining working gizmos on test r_2 were removed before the end of test duration (72 hrs.) to satisfy customer demands. The rest were still working at the end of the 72-hr. test.

(4) *Leukemia patients*: Leukemia is the cancer of blood and as in any other type of cancer, there are remission periods. In a remission period, the patient though not free of disease is free of symptoms. The length of the remission period is a variable of interest in this study. The patients in the state of remission are followed over time to see how long they stay in remission.

(5) *A prospective study of heart condition.* A disease-free cohort of individuals is followed over several years to see who develops heart disease and when does it happen.

(6) *Recidivism study*: A recidivist is a person who relapses into crime. In this study, newly released parolees are followed in time to see whether and when they get rearrested for another crime.

(7) *Spring testing*: Springs are tested under cycles of repeated loading and failure time is the number of cycles leading to failure. Samples of springs are allocated to different stress levels to study the relationship between the lifetimes at different stress levels. At the lower stress levels failure times could be longer than at higher stress levels.

Measurement of Survival Time (or Failure Time): The following points should be kept in mind while measuring the survival time. The time origin should be precisely defined for each individual. The individuals under study should be as similar as possible at their time origin. The time origin need *not* be and usually is *not* the same calender time for each individual. Most clinical trials have staggered entries, so that patients enter the study over a period of time. The survival time of a patient is measured from his/her own date of entry. Figures 1.1 and 1.2 show staggered entries and how these are aligned to have a common origin.

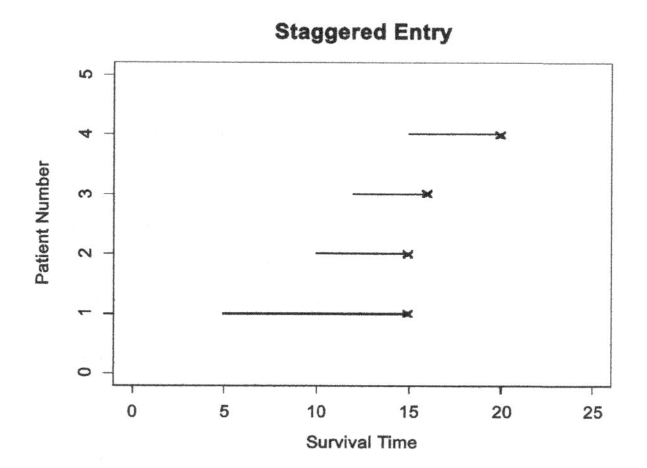

Figure 1.1

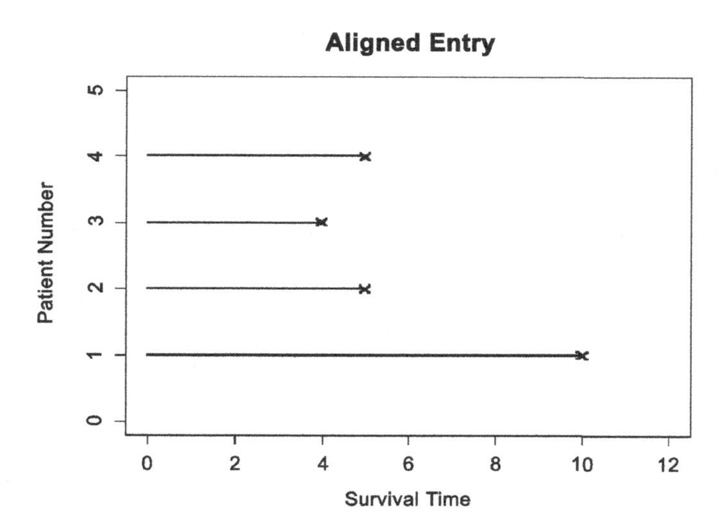

Figure 1.2

The concept of the point event of failure should be defined precisely. If a light bulb, for example, is operating continuously, then the number of hours for which it burned should be used as the lifetime. If the light bulb is turned on and off, as most are, the meaning of number of hours burned will be different as the shocks of lighting and putting off decreases the light bulb's life. This example indicates that there may be more to defining a lifetime than just the amount of time spent under operation.

Censoring: The techniques for reducing experimental time are known as censoring. In survival analysis the observations are lifetimes which can be indefinitely long. So quite often the experiment is so designed that the time required for collecting the data is reduced to manageable levels.

Two types of censoring are built into the design of the experiment to reduce the time taken for completing the study.

Type I (Time Censoring): A number (say n) of identical items are simultaneously put into operation. However, the study is discontinued at a predetermined time t_0. Suppose n_u items have failed by this time and the remaining $n_c = n - n_u$ items remain operative. These are called the censored items. Therefore the data consists of the lifetimes of the n_u failed items and the censoring time t_0 for the remaining n_c items (see Figure 1.3).

Example of type I censoring

Power supplies are major units for most electronic products. Suppose a manufacturer conducts a reliability test in which 15 power supplies are operated over the same duration. The manufacturer decides to terminate the test after 80000 hrs. Suppose 10 power supplies fail during the fixed time interval. Then the remaining five are type I censored.

Type II (Order Censoring): Again a number (say n) of identical components are simultaneously put into operation. The study is discontinued when a predetermined number $k(< n)$ of the items fail. Hence the failure times of the k failed items are available. These are the k smallest order statistics of the complete random sample. For the remaining items the censoring time $x_{(k)}$, which is the failure time of the item failing last, is available (see Figure 1.4).

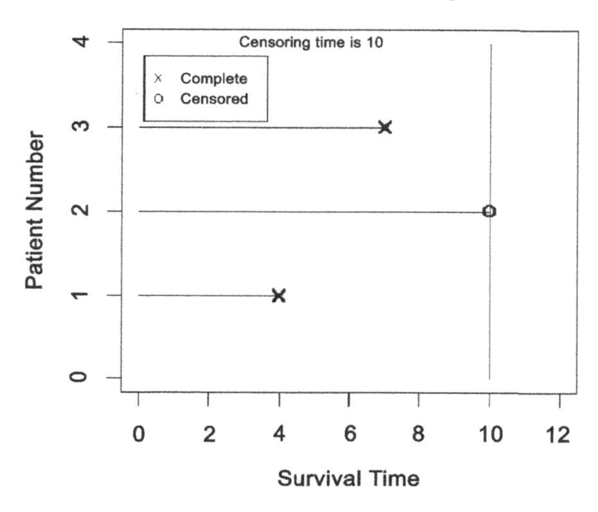

Figure 1.3

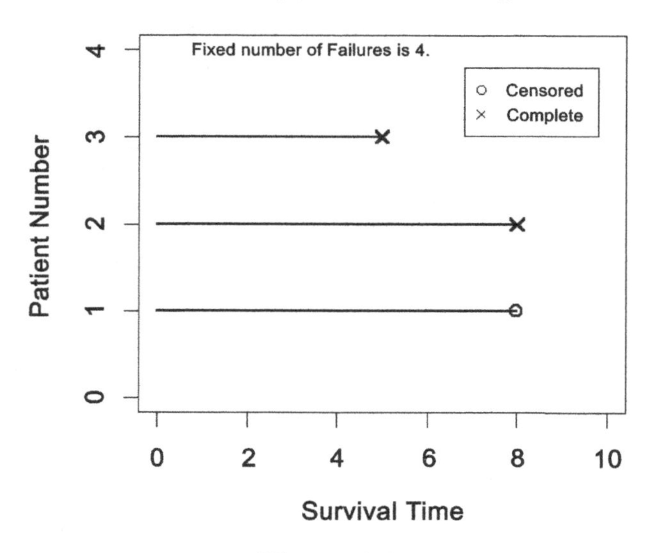

Figure 1.4

Example of Type II censoring

Twelve ceramic capacitors are subjected to a life test. In order to reduce the test time, the test is terminated after eight capacitors fail. The remaining are type II censored.

The above types of censoring are more prevalent in reliability studies (of engineering systems). In survival studies (of biomedical items) censoring is more a part of the experimental situation rather than a matter of deliberate design.

Undesigned censoring occurs when some information about individual survival time is available but exact survival time is *not* known. As a simple example of such undesigned censoring, consider leukemia patients who are followed until they go out of remission. If for a given patient, the study ends while the patient is still in remission (that is the event defining failure does not occur), then the patient's survival time is considered as censored. For this person, it is known that the survival time is not less than the period for which the person was observed. However, the complete survival time is *not* known.

The most frequent type of censoring is known as *right random censoring.* It occurs when the complete lifetimes are not observed for reasons which are beyond the control of the experimenter. For example, it may occur in any of the following situations: (i) loss to follow-up; the patient may decide to move elsewhere and therefore the experimenter may not see him/her again, (ii) withdrawal from the study; the therapy may have bad side effects so it may become necessary to discontinue the treatment or the patient may become non-cooperative, (iii) termination of the study; a person does not experience the event before the study ends, (iv) the value yielded by the unit under study may be outside the range of the measuring instrument, etc. Figure 1.5 illustrates a possible trial in which random censoring occurs. In this figure, patient 1 entered the study at $t = 0$ and died at $T = 5$, giving an uncensored observation. Patient 2 also entered the study at $t = 0$ and was still alive by the end of study, thus, giving a censored observation. Patient 3 has entered the study at $t = 0$ but did not follow up before the end of study to give another censored observation.

Example of Random (right) Censoring

A mining company owns a 1,400 car fleet of 80-ton high-side, rotary-dump gondolas. A car will accumulate about 100,000 miles per year. In their travels from mines to a power plant, the cars are subjected to vibrations due to track input in addition to the dynamic effects of the longitudinal shocks coming through the couplers. As a consequence, the couplers encounter high dynamic impacts and experience fatigue failure and wear. Twenty-eight cars are observed, and the miles driven until the coupler is broken are recroded. The remaining six cars left service after 151000, 155000, 160000, 168000, 175000 and 178000 miles. None of them experienced a broken coupler. Thus giving randomly right censored data.

It may be noted that in type I censoring the number of failures is a random variable whereas in type II censoring the time interval over which the observations are taken is a random variable. In random censoring, the number of complete (uncensored) observations is random and time for which the study lasts may also be random. The censoring time for every censored observation in type I and II censoring is identical, but not so in random censoring. Furthermore, type I censoring may be seen to be a particular case of random censoring by taking all censoring times equal to t_0.

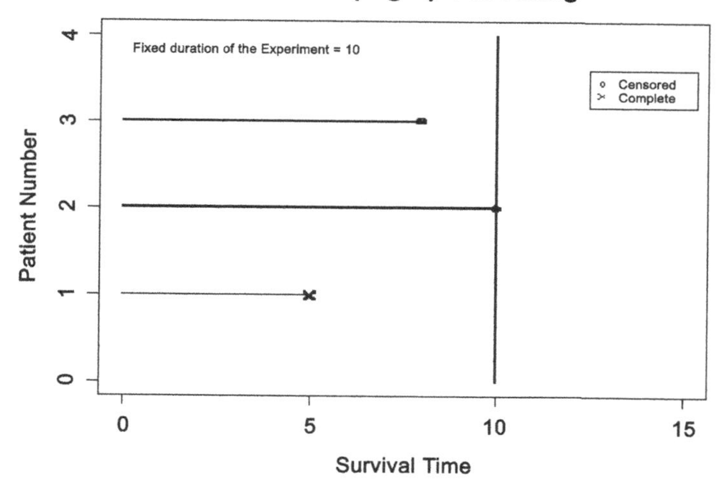

Figure 1.5

Left censoring occurs less frequently than right censoring. It occurs when the observation (time for occurrence of the event) does not get recorded unless it is larger than a certain threshold which may or may not be identical for all observations. For example, the presence of certain gas cannot be measured unless it equals a threshold of six parts per million with a particular measuring device. Such data set will yield left-censored observations.

The data set may contain both left and right censored observations. A psychiatrist collected data to determine the age at which children have learned to perform a particular task. The lifetime was the time the child has taken to learn to perform the task from date of birth. Those children who already knew how to perform the task, when he arrived at the village were left censored and those who did not learn the task even by the time he departed were right-censored observations.

Interval censoring is still another type of censoring for which life time is known only to fall into an interval. This pattern occurs when the items are checked periodically for failure, when a recording instrument has lower as well as upper bounds on its measuring capacity etc.

A simple minded approach to handling the problem of censoring is to ignore all censored values and to perform analysis only on those items that were observed to fail. However, this is not a valid approach. If, for example, this approach is used for right censored data, an overly pessimistic result concerning the mean of the lifetime distribution will result since the longer lifetimes were excluded from the analysis. The proper approach is to provide probabilistic models for the censoring mechanism also.

The second chapter entitled 'Ageing' actually is concerning the development of various mathematical models for the random variable of lifetime. We assume it to be a continuous, positive valued random variable. We make a case for the exponential distribution as the central probability distribution, rather than the normal distribution which is accorded this prime position in standard statistical theory. We discuss various properties of the exponential distribution. The notions of no-ageing and ageing rightly act as indicators while choosing the appropriate law. Positive ageing describes in many ways the phenomenon, that a unit which has already worked for some time has less residual lifetime left than a similar new unit, whereas negative ageing describes the opposite notion. This chapter concerns with many such weak and strong notions and defines nonparametric classes of probability distributions characterised by them. Starting with the exponential distribution as the sole no-ageing distribution, we go on to define Increasing Failure Rate (IFR), Increasing Failure Rate Average (IFRA) and larger classes of distributions and their duals and discuss the properties of these classes. We also introduce the notion of a coherent system of components and show that the lifetime of such a system tends to have a distribution belonging to the IFRA class under fairly general positive ageing conditions on the components. We round off this chapter by providing certain bounds for the unknown distributions belonging to the IFRA class in terms of the exponential distribution with the same value of a moment or of a quantile.

In Chapter 3 we go on to discuss many parametric families of probability distributions which are of special interest in life studies due to their ageing properties. These include direct generalisations of the exponential distribution such as the Weibull and the gamma families as well as Pareto and lognormal. We discuss the ageing and other properties and conclude with some notes on heuristic choice of a family for the experiment under consideration.

The fourth chapter deals with inference for the parameters of the distributions introduced in the previous chapter. We adopt the standard likelihood based frequentist inference procedures. As is well known, except for a few parametric models, the likelihood equations do not yield closed form solutions. In such cases one needs to obtain numerical solutions. These procedures are also described.

As explained above a distinguishing feature of data on lifetimes is the possibility of censored observations, either due to design or necessity. The realisation that censored observations too are informative and should not be discarded is often seen by many as the true beginning of life data analysis. In the fourth chapter we present the modifications required in standard inference procedures in order to take care of censored data as well.

Beginning with the fourth chapter a distinctive feature of the book makes its appearance. It is data analysis on personal computers using R, a software system for statistical analysis and graphics created in the last decade. An introduction to R including reasons for its suitability and adoption are provided in the Appendix at the end of the book. In the fourth chapter we present the commands required for parametric analysis of data arising from exponential and other common life distributions.

In the fifth chapter we introduce nonparametric methods. The first problem to be tackled is that of estimating the distribution and the survival functions. Beginning with the empirical distribution function in the classical setting of complete observations, we go on to the Kaplan-Meier estimator to be used in the presence of randomly censored observations. In such functional estimation one has to appeal to methods of weak convergence or martingale and other stochastic processes based convergence. We therefore provide only indications of proofs of some results. We conclude the chapter with illustrations of R-based computations of the estimates and their standard errors.

The sixth chapter deals with tests of goodness of fit of the exponential distribution. In the context of life data it is important to decide whether the exponential model is appropriate, and if not, the direction of the possibly true alternative hypothesis. Since most of the statistics used for these tests are U-statistics (in the sense of Hoeffding) we devote the second section in this chapter to its development. A more complete development of U-statistics may be found in books on Nonparametric Inference. Besides a number of analytic tests for exponentiality we also introduce certain graphical procedures based on the total time on test (TTT) transform. As in earlier chapters we illustrate these techniques through the R-software.

Next in the seventh chapter we deal with two sample nonparametric methods. We begin with an introductory section on two sample U-statistics and go on to discuss several tests for this problem. These include the Wilcoxon-Mann-Whitney (W-M-W) tests for location differences for complete samples. We discuss Gehan's modification of the W-M-W test for censored samples and further the Mantel-Haenszel, Tarone-Ware classes of statistics and the long-rank test of Peto and Peto. It is our experience that the Kaplan-Meier estimation of the survival function and the Mantel-Haenszel two sample tests are the two most frequently included methods of life data analysis in general statistical softwares. In this as in the previous chapter we provide the R-commands for the application of these two procedures.

We proceed to regression problems in Chapter 8. In classical statistical inference regression is discussed as the effect of covariates on the means of the random variables, or in the case of dichotomous variables, on the log odds ratio. Cox (1972), in a path breaking contribution, suggested that the effects of the covariates on the failure rate are relevant in lifetime studies. He developed a particularly appealing and easy to administer model in terms of effects of covariates which are independent of the age of the subject. This model is called the proportional hazards model. It is a semiparametric model in terms of a baseline hazard rate (which may be known or unknown) and a link function of regression parameters which connect the values of the covariates to it. We provide standard methodology for estimating these parameters and testing hypotheses regarding them in case of complete or censored observations. There are two more models, which are relevant in lifetime studies, which bring in the effects of covariates on the output. The first one is the 'Frailty model' which regards one or more of the explanatory covariates as an unobservable random variable. Assuming a suitable probability distribution for this explanatory random variable, the analysis is carried out. The second such model is the 'Accelerated Lifetime' model (or ALT model). Here the possibility of conducting experiments conditions harsher than those existing in actual environment. This is done to enable faster experimentation. The connection between the actual model and the model appropriate for the harsher experimental conditions is through the introduction of certain covariates. Such analysis is also described in this chapter. R-commands to carry out these procedures are also provided.

The ninth chapter too considers a problem which was first considered in the setting of lifetime studies. The failure of a unit, when it occurs, may be ascribed to one of many competing risks. Hence the competing risks data consists of (T, δ), the time to failure (T) as well as the cause of failure (δ). Later on this model was extended to any situation which looked at the time of occurrence of a multinomial event along with the event that occured, as the basic data. We discuss both parametric and nonparametric methodology for such data, pointing out the non-identifiability difficulties which arise in the case of dependent risks.

So far we have discussed the problems of statistical inference in the classical setting of a random sample consisting of independent and identically distributed random variables, sometimes subject to censoring. In the tenth chapter we consider repairable systems which upon failure are repaired and made operational once more. The data is then in the form of a stochastic process. The degree of repair is an issue. We consider the minimal repair discipline which specifies that the system after repair is restored to the operational state and is equivalent to what it was just prior to failure. The nonhomogeneous Poisson process (NHPP) is seen to be an appropriate model in this context. We discuss estimation of parameters as well as certain tests in this context in this chapter. Certain R-commands are provided for fitting a piecewise constant intensity function to such data.

The first edition of the book did not contain many exercises to be worked out by students. In this edition we include many such exercises which will allow the students to practise the set of techniques learnt here.

We conclude the book with an Appendix which introduces statistical analysis using R. The basic methodology including its installation, methods of data input, carrying out the required analysis and other necessary information is provided here.

Chapter 2

Ageing

2.1 Introduction

The concept of ageing plays an important role in the choice of models for the lifetime distributions. It is particularly useful in engineering applications to model the lifetime distributions of units subjected to wear and tear or shocks.

Let T be a continuous, non-negative valued random variable representing the lifetime of a unit. This is the time for which an individual (or unit) carries out its appointed task satisfactorily and passes into "failed" or "dead" state thereafter. The age of the working unit or living individual is the time for which it is already working satisfactorily without failure. No states besides "living" (operating) or "dead" (failed) are envisaged.

2.2 Functions Characterising Lifetime Random Variable

The probabilistic propeties of the random variable are studied through its cumulative distribution function F or other equivalent functions defined below:

(i) *Survival function or Reliability function*

$$\overline{F}(t) = 1 - F(t) = P[T > t], \quad t \geq 0.$$

(ii) *Probability density function*

$$f(t) = \frac{d}{dt}F(t) = -\frac{d}{dt}\overline{F}(t),$$

(when it exists).

(iii) *Hazard function or failure rate function*

$$\begin{aligned} r(t) &= \lim_{0<h\to 0} \frac{1}{h} P[t < T \leq t+h | T > t] \\ &= \lim_{0<h\to 0} \frac{\overline{F}(t) - \overline{F}(t+h)}{h\overline{F}(t)} \\ &= \frac{f(t)}{\overline{F}(t)}, \text{ provided } F(t) < 1 \text{ and } f(t) \text{ exists.} \end{aligned}$$

Conversely,

$$\overline{F}(t) = \exp\{-\int_0^t r(u)du\}.$$

(iv) *Cumulative hazard function*

$$R(t) = \int_0^t r(u)du, \quad t \geq 0.$$

Therefore

$$\overline{F}(t) = \exp[-R(t)].$$

(v) *Mean Residual life function*

Let a unit be of age t. That is, it has survived without failure upto time t. Since the unit has *not* yet failed, it has certain amount of residual lifetime. Let T_t be the residual lifetime and $\overline{F}_t$ be its survival function.

$$\overline{F}_t(x) = P[T_t > x] = P[T > t+x | T > t] = \frac{\overline{F}(t+x)}{\overline{F}(t)}.$$

Then the mean residual life function is defined as

$$L_F(t) = E[T_t] = \int_0^\infty \overline{F}_t(u)du = \int_0^\infty \frac{\overline{F}(t+u)}{\overline{F}(t)}du, \quad t \geq 0.$$

This gives,

$$L_F(0) = E(T) = \mu$$

and

$$r(t) = [1 + L'(t)]/L(t).$$

(vi) *Equilibrium distribution function*

Suppose identical units are put into operation consecutively, i.e. a new unit is put in operation immediately after the failure of the one in operation.

The lifetimes of these units are assumed to be independent identically distributed random variables (i.i.d.r.v.s), with distribution function F. Let us consider the residual lifetime of a unit in operation at time t as $t \to \infty$. The distribution function of this lifetime is called the equilibrium distribution function, say H_F. From renewal theory we have

$$H_F(t) = \frac{1}{\mu}\int_0^t \overline{F}(u)du, \quad \mu = E(T) = \int_0^\infty \overline{F}(u)du.$$

It can be verified that H_F is a proper distribution function. Let

$$\begin{aligned} r_H(t) &= \text{ failure rate of equilibrium distribution} \\ &= \frac{\overline{F}(t)}{\overline{H}_F(t)}.\frac{1}{\mu}. \end{aligned}$$

Then

$$r_H(0) = \frac{1}{\mu}$$

and

$$\overline{F}(t) = \frac{r_H(t)}{r_H(0)} \exp\{-\int_0^t r_H(u)du\}.$$

One-to-one correspondance of all the above functions is clearly seen. A modeller uses the function which brings out the interesting properties most clearly.

2.3 Exponential Distribution as Model for No-Ageing

Let a unit be of age t. It has residual lifetime T_t with $\overline{F}_t$ as its survival function.

The unit does not aged at all or age has no effect on the residual lifetime of the unit or a used unit of age t (for all t) is as good as a new unit, are all descriptions of the no-ageing phenomenon.

(a) A mathematical way to describe it would be to say that $T_t (t \geq 0)$ are identically distributed random variables. That is,

$$\overline{F}(x) = \overline{F}_t(x) \quad \forall \ t, x \geq 0.$$

Or

$$\overline{F}(x) = \frac{\overline{F}(t+x)}{\overline{F}(t)} \quad \forall \ t, \ \ x \geq 0.$$

Or

$$\overline{F}(x)\overline{F}(t) = \overline{F}(t+x).$$

The last equation is the celebrated Cauchy functional equation. It is well known that among the continuous distributions only the exponential distribution, $\overline{F}(t) = e^{-\lambda t}, t > 0$ satisfies it. This characteristic property of the exponential distribution is also called "lack of memory" property. In lifetime studies we refer to this property as the no ageing property.

Theorem 2.3.1: $\overline{F}(t+x) = \overline{F}(t)\overline{F}(x)$ iff F is the distribution function of the exponential distribution.

Proof. (of the only if part).

We know that

$$\overline{F}(t+x) = \overline{F}(t)\overline{F}(x).$$

Then

$$\overline{F}(nc) = (\overline{F}(c))^n$$

and

$$\overline{F}(c) = (\overline{F}(c/m))^m.$$

Claim: $0 < \overline{F}(1) < 1$. For, (i) if $\overline{F}(1) = 1$ then $\overline{F}(n) = (\overline{F}(1))^n$.

Therefore, $\lim_{n\to\infty} \overline{F}(n) = \overline{F}(\infty) = 1$. This is a contradiction.

Hence $\overline{F}(1) < 1$. (ii) If $\overline{F}(1) = 0 \Rightarrow \lim_{m\to\infty} \overline{F}(1/m) = 0 \Rightarrow \overline{F}(0) = 0$ which again is a contradiction.

$$\begin{aligned} \text{Let } \overline{F}(1) &= e^{-\lambda}, \;\; 0 < \lambda < \infty. \\ \text{Then } \overline{F}(1/m) &= e^{-\lambda/m}, \\ \text{and } \overline{F}(n/m) &= e^{-\lambda n/m}. \\ \text{Therefore } \overline{F}(y) &= e^{-\lambda y} \text{ for all rational } y > 0. \end{aligned}$$

The set of rationals is dense in $\mathcal{R}$ and $\overline{F}$ is continuous. Hence $\overline{F}(y) = e^{-\lambda y}$ for all $y > 0$.

(b) For the exponential distribution, $r(t) = \lambda$; that is, the failure rate is constant. This characterisation of the exponential distribution also expreses its no ageing property.

(c) Consider the mean residual life function:

$$L_F(t) = \mu_F(t) = \int_0^\infty \overline{F}_t(x)dx.$$

For exponential distribution:

$$\mu_F(t) = \int_0^\infty \overline{F}(x)dx$$
$$\Leftrightarrow \mu_F(0) = \mu \;\; \forall \;\; t > 0.$$

That is, the exponential distribution or no ageing is characterised by constant mean residual life also.

(d) Yet another characterisation of interest of the exponential distribution is in terms of its equilibrium distribution, defined as

$$H_F(t) = \frac{1}{\mu}\int_0^t \overline{F}(u)du, \quad \mu = E(T).$$

Let G be the exponential distribution function.

$$\begin{aligned} H_G(t) &= \frac{1}{\mu}\int_0^t e^{-\lambda u}du \text{ where } \quad \mu = \frac{1}{\lambda} \\ &= 1 - e^{-\lambda t}, \;\; t \geq 0 \\ &= G(t). \end{aligned}$$

Similarly, the converse may be proved. Therefore no ageing is equivalent to

$$H_F(t) \equiv F(t).$$

(e) Define

$$\Psi_F(t) = \frac{1}{\mu}\int_0^{F^{-1}(t)} \overline{F}(u)du, \;\; 0 \leq t \leq 1 \text{ and } \quad \mu = E(T).$$

Ψ_F is known as the scaled "total time on test" (TTT) transform of F provided F^{-1} exists and is unique. Trivially $H_F(t) = \Psi_F(F(t))$.

No ageing or the exponential distribution is characterised by

$$\Psi_F(t) = t, \;\; 0 \leq t \leq 1,$$

for

$$\Psi_F(t) = \lambda \int_0^{F^{-1}(t)} e^{-\lambda u}du = t, \;\; 0 \leq t \leq 1.$$

In short, NO AGEING can be described as
(i) Cauchy functional equation
(ii) Constant failure rate
(iii) Constant mean residual life

(iv) Exponential life distribution
(v) Exponential equilibrium distribution.
(vi) Identity function as the TTT transform,
and through many other concepts.

Electronic items, light bulbs etc., often exhibit the "no ageing" phenomenon. These items do not change properties with usage, but they fail when some external shock like a surge of high voltage, comes along. It can be shown that if these shocks occur according to a Poisson process then the lifetime of the item has exponential distribution.

Practical implications of no ageing:

(a) Since a used component is as good as new (stochastically), there is no advantage in following a policy of planned replacement of used components known to be still functioning.

(b) In statistical estimation of mean life, percentiles, survival function etc. the data may be collected consisting only of observed lifetimes and the number of observed failures; the ages of the components under observation are irrelevant.

In what follows, we shall see how the departures from the characterisations of no ageing, in specific directions describe various kinds of ageing properties.

2.4 Positive Ageing

The no ageing situation is adequately described by the exponential distribution. In fact, it is the only possible model for the lifetime of a non-ageing unit. However, in real life, the positive ageing phenomenon is observed quite often. By positive ageing we mean that the age has adverse effect on the residual lifetime of the unit. We shall describe various ways of modelling positive ageing. The different ways of describing negative ageing can then be obtained from the positive ageing descriptions by making appropriate changes.

(i) Increasing Failure Rate (IFR) class of distributions: We shall first define the concept of stochastic dominance. If X and Y are the two random variables then X is "stochastically smaller" than $Y (X \overset{st}{\leq} Y)$ if $F(x) \geq G(x) \;\; \forall \;\; x$, where F and G are distribution functions of X and Y respectively. Obviously

$$F(x) \geq G(x), \quad \forall x \Leftrightarrow \overline{F}(x) \leq \overline{G}(x), \quad \forall x.$$

That is, $P[X > x] \leq P[Y > x]$, $\forall\ x$. Therefore, r.v. Y takes values greater than x with larger probability than the r.v. X for any given real x. Hence Y is stochastically larger than X or Y is said to dominate X stochastically.

We shall investigate the effects of ageing on the performance of the units in terms of stochastic dominance.

If age affects the performance adversely, i.e. the residual lifetime of unit of age t_2 is stochastically shorter than residual lifetime of a unit of age $t_1 (t_1 < t_2)$ then that could be stated as

$$X_{t_1} \overset{st}{\geq} X_{t_2} \ \forall\ 0 \leq t_1 \leq t_2.$$

Or equivalently, in terms of survival functions:

$$\begin{aligned}
\overline{F}_{t_1}(x) \geq \overline{F}_{t_2}(x) \quad & \forall\ 0 \leq t_1 \leq t_2 \\
\Leftrightarrow \overline{F}_t \downarrow t & \qquad (2.4.1) \\
\Leftrightarrow \lim_{x \to 0} \frac{1}{x}(1 - \overline{F}_t(x)) \uparrow t & \\
\Leftrightarrow \lim_{x \to 0} \frac{1}{x}[1 - \frac{\overline{F}(t+x)}{\overline{F}(t)}] \uparrow t & \\
\Leftrightarrow \lim_{x \to 0} \frac{1}{x}[\frac{F(t+x) - F(t)}{\overline{F}(t)}] \uparrow t & \\
\Leftrightarrow r(t) \uparrow t, \text{ provided the pdf exists.} & \qquad (2.4.2)
\end{aligned}$$

Thus, the class of distributions known as the increasing failure rate (IFR) class is also exactly the class of distributions F such that (2.4.1) is satisfied $\forall\ x \in \mathcal{R}$. However, it may be noted that (2.4.1) does not require the existence of a density whereas (2.4.2) does. It may also be noted that equality in (2.4.1) or constancy in (2.4.2) means exponential distribution.

The shape of the hazard function indicates how an item ages. The intuitive interpretation of hazard function as the amount of risk an item is subject to at time t, indicates that when the hazard function is large the item is under greater risk than when it is small. The hazard function being increasing means that items are more likely to fail as time passes. In other words, items wear out or degrade with time. This is almost certainly the case with mechanical items that undergo wear or fatigue. It can also be the case in certain biomedical experiments. If T is time until a tumour appears after the carcinogenic injection in an animal experiment, then the carcinogen makes the tumour more likely to appear as time passes. Hence the hazard function associated with T is increasing.

(ii) Increasing Failure Rate Average (IFRA) Class of life distributions:

The failure rate average function is defined as

$$\overline{R}_F(t) = \frac{1}{t}R(t)$$
$$\overline{R}_F(t) = -\frac{1}{t}\log\overline{F}(t).$$

If the function $\overline{R}_F(t)$ is increasing, then the distribution F is said to possess the increasing failure rate average property and is said to belong to the IFRA class.

Characterisation of IFRA distribution

A distribution F is IFRA if and only if

$$\overline{F}(\alpha t) \geq [\overline{F}(t)]^\alpha \text{ for } 0 < \alpha \leq 1 \ \& \ t \geq 0.$$

Proof: F is IFRA

$$\Leftrightarrow \frac{1}{t}\int_0^t r_F(u)du \uparrow t$$
$$\Leftrightarrow -\frac{1}{t}\log\overline{F}(t) \uparrow t$$
$$\Leftrightarrow \overline{F}(t)^{1/t} \downarrow t$$
$$\Leftrightarrow [\overline{F}(\alpha t)]^{1/\alpha t} \geq [\overline{F}(t)]^{1/t} \ \forall t \geq 0 \text{ and } 0 < \alpha \leq 1$$
$$\Leftrightarrow [\overline{F}(\alpha t)] \geq [\overline{F}(t)]^\alpha \ \forall t > 0, \ 0 < \alpha \leq 1.$$

It is obvious that IFR $\Rightarrow$ IFRA as the average of an increasing function is increasing.

Remark: The classes IFR and IFRA are classes of progressive ageing. We shall now consider a weaker form of ageing which is different from progressive ageing.

(iii) New Better than Used (NBU) Class of ageing.

We compare the distribution of the lifetime of a new unit (i.e. r.v. X) with the lifetime of a unit of age $t(> 0)$ [i.e. r.v. X_t]. The distribution function of these two random variables are F and F_t respectively. F is said to have the "New Better than Used" property if

$$\overline{F}(x) \geq \overline{F}_t(x), \ \forall \ x, t > 0.$$

That is,

$$\overline{F}(x)\overline{F}(t) \geq \overline{F}(x+t), \ \forall \ x, t > 0.$$

This is a weaker form of ageing since one does not compare the units at all ages in this criterion.

IFRA $\Rightarrow$ NBU.

For,

$$\begin{aligned} F \text{ is } IFRA &\Leftrightarrow \overline{F}(\alpha t) \geq [\overline{F}(t)]^{\alpha}, t > 0, 0 < \alpha < 1 \\ &\Rightarrow \overline{F}((1-\alpha)t) \geq [\overline{F}(t)]^{1-\alpha} \ \forall \ t > 0, 0 < \alpha < 1 \ \text{ and} \\ &\quad \overline{F}(\alpha t)\overline{F}[(1-\alpha)t] \geq \overline{F}(t) \ \forall \ t > 0, 0 < \alpha < 1 \end{aligned}$$

(Let $t = x + y$ and $\alpha = \frac{x}{x+y}$)

$\Rightarrow \overline{F}(x)\overline{F}(y) \geq \overline{F}(x+y)$

$\Rightarrow F$ is NBU.

(iv) New Better than Used in Expectation (NBUE) Class

A still weaker form of positive ageing is NBUE defined by the inequality

$$\int_0^\infty \overline{F}(x)dx \geq \int_0^\infty \overline{F}_t(x)dx.$$

Or

$$L_F(0) \geq L_F(t) \quad \forall \quad t > 0.$$

It is obvious that NBU $\Rightarrow$ NBUE.

It may be noted that for progressive ageing classes the comparison between units of different ages is possible. However, for NBU and NBUE classes the comparison is between brand new unit and a unit aged t. We shall now consider two more progressive ageing classes.

(v) Decreasing Mean Residual Life (DMRL) Class

Let $E(X_t)$ denote the mean residual lifetime of a unit of age t. Then one can say that $E(X_t) \downarrow t$ is also a way of describing progressive positive ageing. This is called the "Decreasing Mean Residual Life" (DMRL) property.

(a) IFR property $\Rightarrow$ DMRL property.

For,

$$F \text{ is } IFR \Leftrightarrow \overline{F}_{t_1}(x) \geq \overline{F}_{t_2}(x) \ \forall \ t_1 < t_2.$$

By integration, we get

$$\int_0^\infty \overline{F}_{t_1}(x)dx \geq \int_0^\infty \overline{F}_{t_2}(x)dx, \ \forall \ t_1 < t_2.$$

That is $E(X_t) \downarrow t$.

(b) DMRL $\Rightarrow$ NBUE. This is seen by putting $t_1 = 0$ in the above.

(vi) Harmonically New Better than used in Expectation (HNBUE) Class

A distribution F is said to belong to the HNBUE class if

$$\int_t^\infty \overline{F}(x)dx \leq \mu e^{-t/\mu}, \quad t \geq 0, \tag{2.4.3}$$

where $\mu = E(X) = \int_0^\infty \overline{F}(u)du$.

HNBUE property can be equivalently described as

$$[\frac{1}{t}\int_0^t \frac{1}{L_F(x)}dx]^{-1} \leq \mu \text{ for } t > 0. \tag{2.4.4}$$

Note: Definition (2.4.4) explains why the property is named as HNBUE.

We show below that NBUE $\Rightarrow$ HNBUE

$$\begin{aligned} NBUE &\Leftrightarrow L_F(0) \geq L_F(t) \quad \forall t > 0 \\ &\Leftrightarrow \frac{1}{L_F(0)} \leq \frac{1}{L_F(t)} \\ &\Rightarrow \frac{1}{t}[\int_0^t \frac{1}{L_F(0)}dx]^{-1} \geq [\frac{1}{t}\int_0^t \frac{1}{L_F(x)}dx]^{-1} \\ &\Rightarrow \mu \geq [\frac{1}{t}\int_0^t \frac{1}{L_F(x)}dx]^{-1}. \end{aligned}$$

The total picture of implications is shown in Figure 2.1.

Chain of Implications (Positive Ageing)

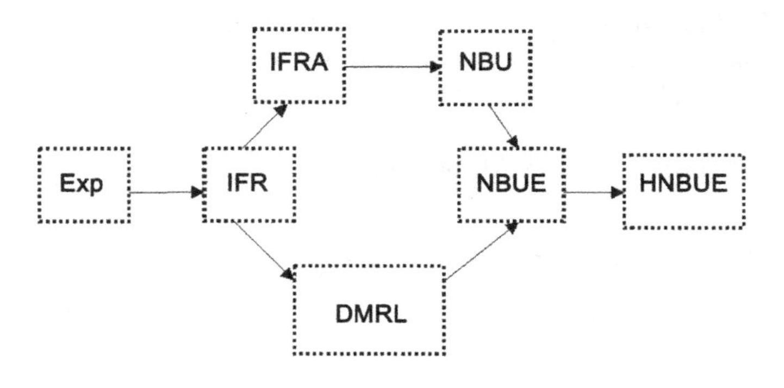

Figure 2.1

2.5 Negative Ageing

For the sake of completeness we also mention similar concepts of beneficial types of ageing (negative ageing). These can be summarised in the following implication chain.

Chain of Implications (Negative Ageing)

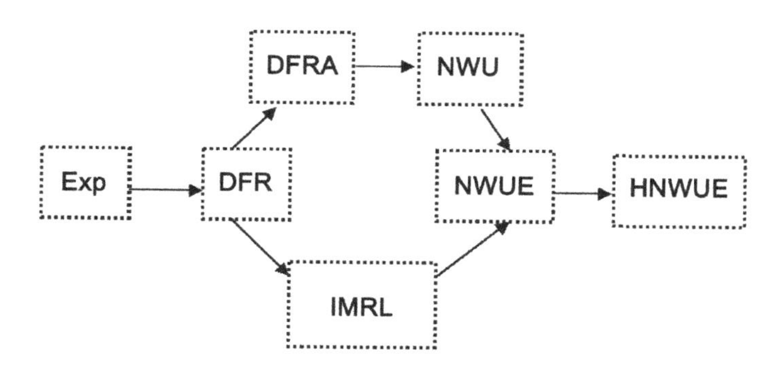

Figure 2.2

In the above diagram DFR stands for decreasing failure rate and DFRA stands for decreasing failure rate average. The notations NWU, NWUE, IMRL and HNWUE are used respectively for "new worse than used", "new worse than used in expectation", "increasing mean residual life" and harmonically new worse than used in expectation."

Apart from the ageing classes considered above, economists have defined and used some ageing classes which are based on the concept of stochastic dominance of order higher than one.

2.6 Relative Ageing of Two Probability Distributions

Let X and Y be positive-valued r.v.s with distribution functions F and G, survival functions $\overline{F}$ and $\overline{G}$ and cumulative hazard functions $R_F = -\log \overline{F}$ and $R_G = -\log \overline{G}$ respectively. We assume the existence of corresponding densities f and g. Then hazard rates are given by $h_F = \frac{f}{\overline{F}}$ and $h_G = \frac{g}{\overline{G}}$ respectively.

Definition: The r.v. X is said to be ageing faster than Y (written '$X \overset{<}{c} Y$' or '$F \overset{<}{c} G$') if the r.v. $Z = R_G(X)$ has increasing failure rate (IFR) distribution.

It is easy to see that the above definition is equivalent to each of the following three statements.

(i) $X \overset{<}{c} Y$ if and only if $R_F o R_G^{-1}$ is convex on $[0, \infty)$.

(ii) $X \overset{<}{c} Y$ if and only if $R_F(Y)$ has DFR distribution.

(iii) If h_F and h_G exist and $h_G \neq 0$ then $X \overset{<}{c} Y$ if and only if $\frac{h_F}{h_G}$ is a non-decreasing function.

The characteristic property (iii) can be interpreted in terms of relative ageing as follows:

If the two failure rates are such that $\frac{h_F(x)}{h_G(x)}$ is a constant, then one may say that the two probability distributions age at the same rate. On the other hand if the ratio is an increasing (decreasing) function of the age x, then it may be said that the failures according to the random variable X tend to be more and more (less and less) frequent, as age increases as compared to those of Y. Hence, we may say that the distribution of X ages faster (slower) than that of Y.

In survival analysis, we often come across the problem of comparison of treatment abilities to prolong life. In such experiments, the phenomenon of crossing hazard is observed. For example, Pocock et al. (1982) in connection with prognostic studies in the treatment of breast cancer, Champlin et al. (1983) and Begg et al. (1983) in relation to bone marrow transplantation studies, have reported the superiority of a treatment being short lived. In such situations, the hypothesis of increasing (decreasing) hazard ratio will be relevant for the comparison of the two treatments.

There are two simple generalisations of '$\overset{<}{c}$' order which are obtained by replacing 'IFR' in definition by 'IFRA' or 'NBU'.

For details refer to Sengupta and Deshpande (1994).

2.7 Bathtub Failure Rate

Another class of the distributions which arises naturally in human mortality study and in reliability situations is characterised by failure rate functions having "bathtub shape". The failure rate decreases initially. This initial phase is known as the "infant mortality" phase. A good example of this is seen in the standard mortality tables for humans. The risk of death is large for infants but decreases as age advances. Next phase is known as "useful life" phase, in which the failure rate is more or less constant. For example, in human mortality tables it is observed that for the ages 10 - 30 years, the death rate is almost constant at a level less than that for the previous period. The cause of death, in this period, is mainly attributed to accidents. Finally, in the third phase, known as "wearout phase", the failure rate increases. Again in humans, after the age of 30 an increasing proportion of the alive persons die as age advances. The three phases of failure rates are represented by a bathtub curve (see Figure 2.3).

An empirical illustration of a bathtub shaped failure rate reproduced from Barlow and Proschan (1975) (originally from Kamins (1962) is shown in Figure 2.4.

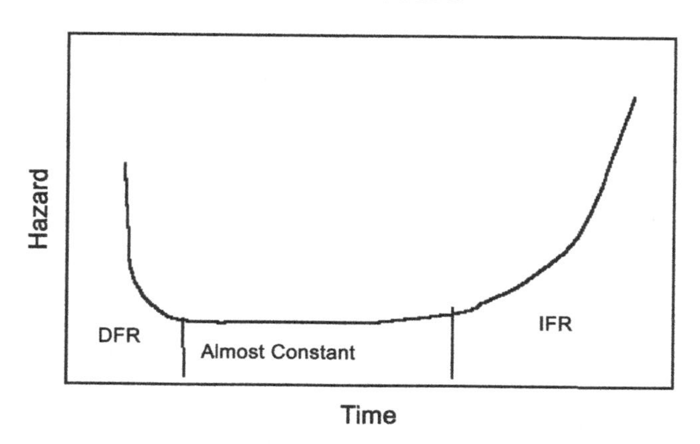

Figure 2.3

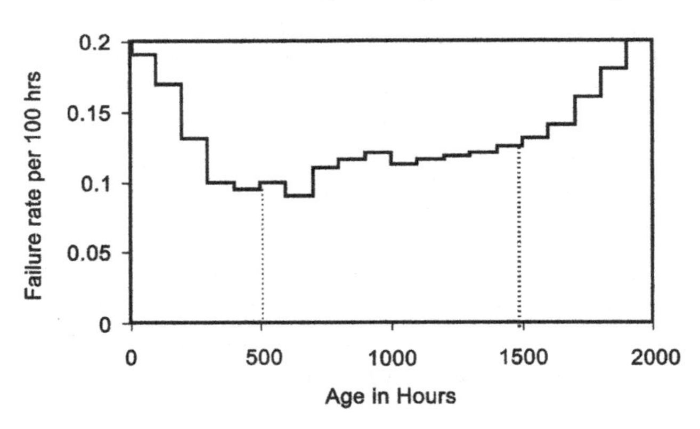

Figure 2.4

The ordinate represents the empirical failure rate per 100 hours for a hot-gas generating system used for starting the engines of a particular commercial airliner. During the first 500 hours of operation, the observed failure rate decreases by about half. From 500 to 1500 hrs of operation, the failure rate remains nearly constant and finally after 1500 hours of operation the failure rate increases.

Remark: We have considered three types of hazard function curves (i) IFR, (ii) DFR and (iii) bathtub. The increasing hazard function is probably the most likely situation of the three. In this case, items are more likely to fail as time passes. The second situation, the decreasing hazard function, is less common. In this case, the item is less likely to fail as time passes. Items with this type of hazard function improve with time. Some metals work harden through use and thus have increased strength as time passes. Another situation for which a decreasing hazard function might be appropriate for modelling, is debugging of computer programs. Bugs are more likely to appear initially, but the chance of them appearing decreases as time passes. The bathtub shaped hazard function can be envisioned in different situations apart from the ones already discussed. Suppose there are two factories which produce the same item. Factory A produces high quality items which are expensive and factory B produces low quality, cheap items. If p_1 and p_2 are respectively probabilities of selection of the items from the two factories then a mixture distribution is the appropriate model for lifetime of an item in the selected lot. In such mixtures of the items decreasing or bathtub hazard rates are possible. Human performance tasks, such as vigilance, monitoring, controlling and tracking are possible candidates for modelling hazards by bathtub curves. In these situations the lifetime is time to first error. The burn-in (or infant mortality) period corresponds to learning and the wear-out period corresponds to fatigue.

2.8 System Lifetime

So far we have discussed random lifetime and its probability distributions for a single unit to be identified with a component. A system on the other hand may be regarded to be composed of many such components. Obviously, the lifetime and the probability distributions for the system as a whole will be based upon those for its components. In order to study these interdependences, we introduce the following notation.

Let a system be composed of n components. Designate $x_i, i = 1, 2, \cdots, n$, binary variables to indicate the state of the n components respectively:

$$x_i = \begin{cases} 1, & \text{if the } i\text{-th component is functioning} \\ 0, & \text{if the } i\text{-th component has failed.} \end{cases}$$

Further, let $\phi(x_1, x_2, \cdots, x_n)$ be the structure function of the system denoting its state

$$\phi(x_1, x_2, \cdots, x_n) = \begin{cases} 1, & \text{if the system is functioning} \\ 0, & \text{if the system has failed.} \end{cases}$$

For example, a series system is one which functions as long as all its components are functioning. Hence its structure function can be specified as $\phi(x_1, x_2, \cdots, x_n) = \prod_{i=1}^{n} x_i$, which is equal to one if and only if all the x_i's are equal to one, i.e. all the components are working and zero otherwise. A parallel system is one which keeps functioning until at least one of its components is functioning. Its structure function can be seen to be

$$\phi(x_1, x_2, \cdots, x_n) = 1 - \prod_{i=1}^{n}(1 - x_i).$$

A k-out-of-n system is one which functions as long as at least k of its n components are functioning.

In reliability theory the IFRA class occupies an important place. It is the smallest class of distributions which contains the exponential distribution and is closed under the formation of "coherent" systems of independent components. Most of the systems observed in practice are coherent systems. We shall now discuss this important class of systems.

Coherent Systems: A system is a coherent system if its structure function satisfies the following two conditions.

(i) *Relevancy of a Component*: There exists some configuration of the states $x_1, x_2, \cdots, x_{i-1}, x_{i+1}, \cdots, x_n$ along with which the state of the i-th component matters to the system. Symbolically,

$$\phi(x_1, \cdots, x_{i-1}, 0, x_{i+1}, \cdots, x_n) = 0$$

and

$$\phi(x_1, \cdots, x_{i-1}, 1, x_{i+1}, \cdots, x_n) = 1$$

for some configuration of $(x_1, \cdots, x_{i-1}, x_{i+1}, \cdots, x_n)$ for every i.

(ii) *Monotonocity of the Structure Function*: If a failed component in a system is replaced by a functioning component, then the state of the system must not change from functioning to failed. Again symbolically,

$$\phi(x_1, \cdots, x_{i-1}, 0, x_{i+1}, \cdots, x_n) \leq \phi(x_1, \cdots, x_{i-1}, 1, x_{i+1}, \cdots, x_n)$$

for every i and for every configuration $(x_1, \cdots, x_{i-1}, x_{i+1}, \cdots, x_n)$.

Henceforth we will assume that we are dealing only with a coherent system. All common systems, such as series, parallel, k-out-of-n, etc. are seen to be coherent.

Let p_i be the probability that the i-th component is functioning at the time of interest. It is also called its reliability. Also, assume that the components function in statistically independent manner. Define $h_\phi(\underline{p})$, where $\underline{p} = (p_1, p_2, \cdots, p_n)$ to be the probability that the system having structure function ϕ works at the time of interest, i.e. it is the reliability of the system. Then it is easily argued that

$$h_\phi(\underline{p}) = \sum_{\underline{x}} \phi(\underline{x}) \prod_{i=1}^{n} p_i^{x_i}(1 - p_i)^{1-x_i}$$

where the summation is over all 2^n vectors $(x_1, x_2, \cdots, x_n)$ of 0's and 1's indicating the state of the components. Examples of system reliabilities include

(i) Series system: $h(\underline{p}) = \prod_{i=1}^{n} p_i$

(ii) k-out-of-n system: $h(\underline{p}) = \sum_{j=k}^{n} \binom{n}{j} p^j(1-p)^{n-j}$ provided $p_1 = \cdots = p_n = p$.

(iii) Parallel system: $h(\underline{p}) = 1 - \prod_{i=1}^{n}(1-p_i)$.

It is also seen that for any arbitrary coherent system the inequality

$$\prod_{i=1}^{n} p_i \leq h_\phi(\underline{p}) \leq 1 - \prod_{i=1}^{n}(1-p_i),$$

holds, i.e. the series system is the weakest and the parallel system is the strongest coherent system or, the least reliable and the most reliable that may be constructed out of these n components. In other words the reliability of an arbitrary coherent system is always with these two bounds.

Illustration 2.1: Two independent components have hazard rates

$$h_1(t) = 1 \text{ and } h_2(t) = 2, \quad t \geq 0.$$

Comment on the ageing properties of the parallel system formed of these.

We find the hazard rates of system time to failure and the mean time to failure of the parallel system of these components.

The survival functions of the two components are $\overline{F}_1(t) = e^{-t}$ and $\overline{F}_2(t) = e^{-2t}$ for $t \geq 0$. The survival function $\overline{F}(t)$ of the two-component parallel system is given by

$$\begin{aligned}\overline{F}(t) &= 1 - (1 - \overline{F}_1(t))(1 - \overline{F}_2(t)) \\ &= e^{-t} + e^{-2t} - e^{-3t}, t \geq 0.\end{aligned}$$

(i) Hazard rate of the system $(h(t))$:

$$h(t) = \frac{-\overline{F}'(t)}{\overline{F}(t)} = \frac{e^{-t} + 2e^{-2t} - 3e^{-3t}}{e^{-t} + e^{-2t} - e^{-3t}}, t \geq 0.$$

(ii) mean time to failure

$$\mu = \int_0^\infty \overline{F}(t)dt = \int_0^\infty (e^{-t} + e^{-2t} - e^{-3t})dt. = 1 + \frac{1}{2} - \frac{1}{3} = \frac{7}{6}.$$

The mean time to failure of the stronger component is 1 and mean time to failure of the weaker component is $\frac{1}{2}$. Thus the addition of the weaker component in parallel with the stronger component increases the mean time to failure by $\frac{1}{6}$.

Each of the two independent components possesses the no ageing property yet the system does not have the no-ageing property. If we plot the hazard rate of this system over time we get hazard curve of Figure 2.5 which is not monotonic.

Cuts and Paths of a Coherent System

A *path* of a coherent system is a subset of its components such that the system works if all the components in this subset function. A *minimal path* of a coherent system is a path but no proper subset of it is a path.

A *cut* of a coherent system is a subset of its component such that the failure of all the components in it leads to the failure of the system. A *minimal cut* is a cut such that no proper subset of it is a cut. Let $P_1, \cdots, P_p$ and $K_1, \cdots, K_k$ denote all the minimal paths and all the k minimal cuts of the coherent system with structure function ϕ. Then the above definitions lead to the following identities:

$$\phi(\underline{x}) = \max_{1 \leq j \leq p} \min_{i \in P_j} x_i = \min_{1 \leq j \leq k} \max_{i \in K_j} x_i.$$

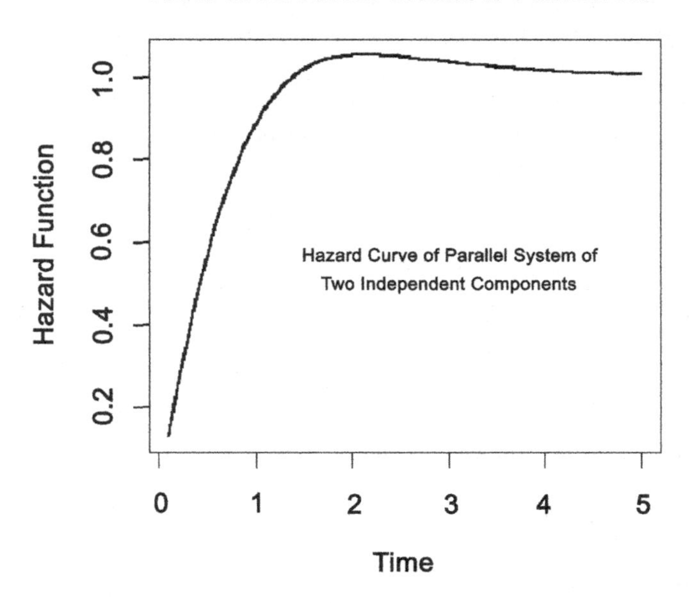

Figure 2.5

In order to gain more insight in the structure function $\phi(\underline{x})$ and reliability function $h(\underline{p})$ of a coherent system, we introduce random variables X_i, indicating the state of the i-th component, $i = 1, 2, \cdots, n$. It is clear that these are independent with probability distribution $P(X_i = 1) = p_i$ and $P(X_i = 0) = 1 - p_i$ and

$$h(\underline{p}) = E[\max_{1 \le j \le p} \min_{i \in P_j} X_i] = E[\min_{1 \le j \le k} \max_{i \in K_j} X_i].$$

Since the cuts or the paths are not necessarily nonoverlapping sets of components, the p minima $\min_{i \in p_j} X_i$ or the k maxima $\max_{i \in k_j} X_i$ are not independent random variables, even if $X_1, X_2, \cdots, X_n$ are. In many reliability situations we come across random variables which are not independent. But usually such variables are "associated". Two random variables S and T may be called associated if $cov[S,T] \ge 0$. A stronger requirement would be $cov(f(S), f(T)) \ge 0$ for all increasing function f and g. Finally, if $cov[f(S,T), g(S,T)] \ge 0$ for all f and g increasing in each argument, we would have a still stronger version of association. This strongest version of association has a natural multivariate generalisation which serves as a definition of association. Using this, the random variable $\min_{i \in P_j} X_i$ or $\max_{i \in K_j} X_i$ are associated. By the properties of associated random variables one can prove the following inequalities:

$$\prod_{j=1}^{k} \{1 - \prod_{i \in K_j} (1 - p_i)\} \le h(\underline{p}) \le 1 - \{\prod_{j=1}^{p} (1 - \prod_{i \in P_j} p_i)\}$$

which utilise the cut and path structure of the structure function $\phi(\underline{x})$, to provide bounds for system reliability.

2.9 IFRA Closure Property

Earlier in this chapter we introduced the IFRA property of life distributions. Here we show that the IFRA class of life distributions is closed under formation of coherent systems of independent components each having an IFRA life distribution. Since the exponential distribution belongs to the IFRA class, it follows that lifetime of a coherent system composed of components with independently, exponentially distributed lifetimes will also belong to the IFRA class. Thus IFRA distributions may be used as models for the lifetimes of a large number of systems.

Let $T_1, T_2, \cdots, T_n$ be the lifetimes of the n components which form the coherent system with structure function $\phi(\underline{x})$, and reliability function $h(\underline{p})$.

Let $T_i, i = 1, 2, \cdots, n$ be independent random variables with c.d.f.s $F_i(t)$, belonging to IFRA class, thus satisfying the property $\overline{F}_i(\alpha t) \geq (\overline{F}_i(t))^\alpha$ for all t and $0 < \alpha \leq 1$. Here $\overline{F}_i$ denotes the survival function $1 - F_i$ corresponding to the c.d.f. F_i. Also, let T be the lifetime of the system with c.d.f. and survival function F and $\overline{F}$ respectively.

Let t be a fixed mission time. The reliability of the i-th component at time t is the probability that it is working at time t, i.e. $\overline{F}_i(t)$, and the corresponding reliability of the system is $\overline{F}(t)$. Replacing p_i by $\overline{F}_i(t)$, the expression for the reliability of the system may be represented as the identity

$$h(\overline{F}_1(t), \overline{F}_2(t), \cdots, \overline{F}_n(t)) = \overline{F}(t).$$

In order to prove that $\overline{F}(t)$ belongs to the IFRA class given that $F_i(t)$'s do, we need to prove that

$$\overline{F}_i(\alpha t) \geq (\overline{F}_i(t))^\alpha, \quad i = 1, 2, \cdots, n \quad \Rightarrow \overline{F}(\alpha t) \geq [(\overline{F}(t)]^\alpha.$$

Below we provide an outline of the proof; the details are given in Barlow and Proschan (1975).

Let $\underline{p}^\alpha = (p_1^\alpha, p_2^\alpha, \cdots, p_n^\alpha)$. Then it is seen that the function $h(\underline{p})$ satisfies the inequality $h(\underline{p}^\alpha) \geq h^\alpha(\underline{p})$, for a coherent system.

First we need

Lemma 2.9.1: Let $0 \leq \alpha \leq 1, 0 \leq \lambda \leq 1, 0 \leq x \leq y$. Then

$$\lambda^\alpha \quad y^\alpha + (1 - \lambda^\alpha)x^\alpha \geq [\lambda y + (1 - \lambda)x]^\alpha.$$

The proof follows once we notice that $f(x) = x^\alpha$, for $x \geq 0$ and $0 \leq \alpha \leq 1$ is a concave function. Then we see that if $n = 1$ then either $h(p) \equiv p$, or $h(p) \equiv 0$ or 1. In each of these cases $h(p^\alpha) = [h(p)]^\alpha$. Then by induction argument and by using the above lemma we prove $h(\underline{p}^\alpha) \geq h^\alpha(\underline{p})$. This inequality is used in the second step below.

We know that h is an increasing function in each of its arguments. Hence,

$$\begin{aligned} \overline{F}(\alpha t) &= h(\overline{F}_1(\alpha t), \cdots, \overline{F}_n(\alpha t)) \\ &\geq h(\overline{F}_1^\alpha(t), \cdots, \overline{F}_n^\alpha(t)) \\ &\geq h^\alpha(\overline{F}_1(t), \cdots, \overline{F}_n(t)) = [\overline{F}(t)]^\alpha, \end{aligned}$$

which shows that the system lifetime has an IFRA distribution. In fact, a stronger result is available. The closure of the IFRA class under construction of coherent systems and under limits in distribution is the IFRA class itself, i.e. symbolically

$$\{IFRA\} = \{IFRA\}^{CS,LD}$$

where the superscripts denote the operations under which the closure is being carried out. It is also seen that the IFRA class is the smallest class of distributions containing the exponential distribution and having the above closure properties. This result brings out the importance of an IFRA distribution as a model for the lifetimes of coherent systems. Particularly useful are the following results which provide bounds on the certain reliability function of a coherent system with the help of exponential distributions which share either a moment (say, the mean) or a quantile (say, the median) with the system lifetime distribution.

2.10 Bounds on the Reliability Function of an IFRA Distribution

The IFRA distribution functions are characterised by the property $\overline{F}(\alpha(x)) \geq \overline{F}^{\alpha}(x), 0 < \alpha \leq 1$, or equivalently, $-\log \overline{F}(\alpha\ x) \leq -\alpha \log \overline{F}(x)$. We know that $-\log \overline{F}(t)$ is the cumulative hazard function corresponding to F. The above characterisation of IFRA distribution in terms of its cumulative hazard function says that the function must be starshaped. (A function $g(x)$ is defined to be starshaped if (i) $g(\alpha\ x) \leq \alpha g(x), 0 < \alpha < 1$, or equivalently (ii) $\frac{1}{x}g(x)$ is increasing in x.) Due to the starshaped nature of the function $-\log \overline{F}(t)$, the following result holds:

If $F(x)$ is an IFRA distribution then $\overline{F}(t) - e^{-\lambda t}$ has at most one change of sign, and if there is one it is from $+$ to $-$.

Now, if $\overline{F}$, an IFRA survival function and $e^{-\alpha t}$, the exponential survival function have the same quantile ξ_p of order p, then $\overline{F}(t) \geq e^{-\alpha t}$, for $0 < t \leq \xi_p$ and $\overline{F}(t) \leq e^{-\alpha t}$, for $\xi_p \leq t < \infty$ where $\alpha = -\frac{1}{\xi_p}\log(1-p)$.

The above bound may be translated as below also.

$$\overline{F}(t) \geq (\overline{F}(a))^{t/a}, \quad 0 \leq t \leq a,$$

and

$$\overline{F}(t) \leq (\overline{F}(a))^{t/a}, \quad a \leq t < \infty.$$

This is so because F is IFRA and $-\frac{1}{t}\log \overline{F}(t)$ is increasing in t. Hence for

$t \leq a$

$$-\frac{1}{t}\log \overline{F}(t) \leq -\frac{1}{a}\log \overline{F}(a)$$

which is the same as $\overline{F}(t) \geq (\overline{F}(a)]^{t/a}$. Similarly, the bound when $t \geq a$ can be derived.

Further, we find bounds on the reliability function of a coherent system composed of independent IFRA components. In notation introduced earlier, we have in terms of means $\mu_1, \cdots, \mu_n$ of the $F_i(t)$

$$\begin{aligned}\overline{F}(t) &= h[\overline{F}_1(t), \cdots, \overline{F}_n(t)] \\ &\geq h(e^{-t/\mu_1}, \cdots, e^{-t/\mu_n}) \quad \text{for} \quad t \leq \min(\mu_1, \cdots, \mu_n) \\ &\leq h(e^{-t/\mu_1}, \cdots, e^{-t/\mu_n}) \quad \text{for} \quad t \geq \max(\mu_1, \cdots, \mu_n)\end{aligned}$$

and

$$\begin{aligned}\overline{F}(t) &= h[\overline{F}_1(t), \cdots, \overline{F}_n(t)] \\ &\geq h[\{F_1(a_1)\}^{t/a_1}, \cdots, \{\overline{F}_n(a_n)\}^{t/a_n}] \quad \text{for} \quad t \leq \min(a_1, a_2, \cdots, a_n) \\ &\leq \quad h[\{\overline{F}_1(a_1)\}^{t/a_1}, \cdots, \{\overline{F}_n(a_n)\}^{t/a_n}] \quad \text{for} \quad t \geq \max(a_1, a_2, \cdots, a_n).\end{aligned}$$

The above leaves a gap from $\min(a_1, \cdots, a_n)$ to $\max(a_1, \cdots, a_n)$ in the bounds. However, if we take $a_1 = a_2 = \cdots = a_n$, then the gap disappears and the two bounds are applicable for $t \leq a$ and $t \geq a$ respectively.

The above bounds (obtained in Chaudhari, Deshpande and Dharmadhikari (1991)) are useful in situations where we know that the c.d.f. of the components or the system belongs to the IFRA class without having a precise knowledge of it. Further, we may know the value of a quantile, or the mean of the distribution, either a priori or through some data. Then one can use the above bounds to make conservative (anticonservative) statements regarding the unknown reliability. It has been found that in several situations these bounds are quite close to the actual values.

Exercises

Exercise 2.1: Consider a system with three independent components in parallel. The probabilities of the three components being operational are 0.9, 0.8, and 0.75. Determine the reliability of the system. If the same three components are joined in series, what will be the reliability of the system?

Exercise 2.2: Two independent components are arranged in series. The lifetimes of the two components have hazard rates:

$$h_1(t) = 1 \text{ and } h_2(t) = 2, \quad t \geq 0.$$

Find the hazard rate of the system.

Exercise 2.3: Write the structure function of a k-out-of-n system. What is the structure function of 2-out-of-3 system?

Exercise 2.4: An airplane has four propellers, two on each wing. The airplane will fly (function) if at-least one propeller on each wing functions. Denote the 4 propellers by components 1, 2, 3, and 4 with 1 and 2 on the left wing and 3,4 on the right wing. Consider for a moment, the plane to consist of two wings, and the wings are arranged in series. Find the structure function assuming that the only cause of failure of the plane is propeller failure.

Exercise 2.5: A parallel-series system consists of m parallel paths. Each path has n units connected in series. If all the units are independent, identical and the reliability of a single unit is p, then what will be the reliability of the system?

Exercise 2.6: A series-parallel system consists of n-subsystems in series with m units in parallel in each subsystem. If all the units are identical, independent and reliability of a single unit is p, then what will be the reliability of the system?

Exercise 2.7: Consider four independent components that are connected in series. Each component has a reliability p. The system fails if two consecutive components fail. This system is referred to as consecutive-2-out-4:F system. Determine the reliability of the system. If $p = 0.95$ then what will be the reliability of the system?

Chapter 3

Some Parametric Families of Probability Distributions

3.1 Introduction

In the last chapter, we studied the exponential distribution and certain non-parametric classes of distributions based on ageing properties of individual items and their systems. We know that the exponential distribution exhibits the no-ageing property and it is the only distribution to do so. In this section, we shall consider some other parametric families of distributions which are used to model lifetimes of the individuals with positive (or negative) ageing properties.

3.2 Weibull Family

The Weibull distribution is a generalisation of the exponential distribution that is appropriate for modelling the lifetimes having constant, strictly increasing (and unbounded) or strictly decreasing hazard functions. It is given by the distribution function

$$F(t) = 1 - e^{-\lambda t^{\gamma}}, \quad t > 0, \quad \lambda > 0, \gamma > 0$$

where λ and γ are both positive valued parameters. It is clear that $\gamma = 1$ gives the exponential distribution with mean $\frac{1}{\lambda}$. Hence this may be viewed as a generalisation of the exponential distribution. It is interesting to look at its failure rate.

$$f(t) = -\frac{d}{dt}\overline{F}(t) = \lambda\gamma t^{\gamma-1}e^{-\lambda t^{\gamma}}, \quad t > 0,$$

$$\text{and} \quad r(t) = \frac{f(t)}{\overline{F}(t)} = \lambda\gamma t^{\gamma-1}, \quad t > 0.$$

As t increases from 0 to $\infty, r(t)$ increases provided $\gamma > 1$ and decreases provided $0 < \gamma < 1$. Thus, this parametric family contains both IFR and DFR probability distributions. Other special case occurs when $\gamma = 2$, commonly known as the Rayleigh distribution, for which the hazard rate is a straight line through the origin with slope 2λ.

The parameter λ is known as the scale parameter. The parameter γ is known as the shape parameter. In this case it is the shape parameter which decides whether the distribution belongs to the IFR class or the DFR class.

Figure 3.1 represents the plots for hazard functions for Weibull distribution with scale parameter equal to 1 and shape parameter equal to 0.5, 1, 1.5 and 2.

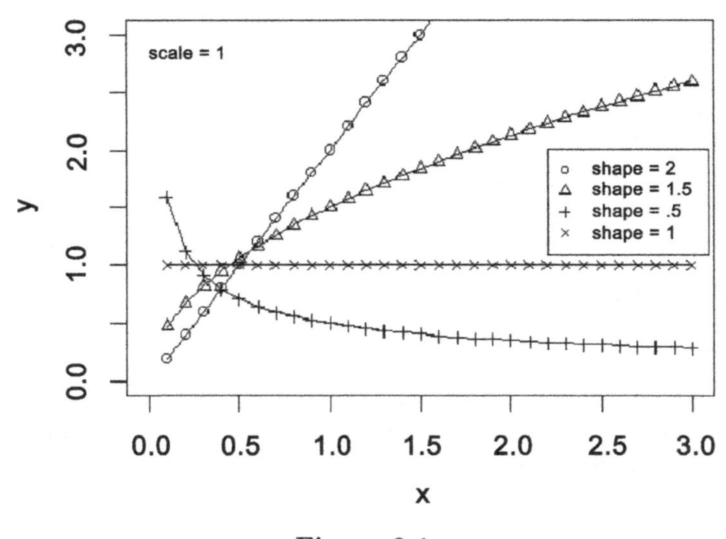

Figure 3.1

When $3 < \gamma < 4$, the probability density function closely resembles that of a normal probability function. The mode and median of the distribution coincide when $\gamma = 3.4393$. The mean residual life function is not as mathematically tractable as the hazard rate. All Weibull survivor functions pass through the point $(1, e^{-\lambda})$, regardless of the value of γ; and since $R(t)$ is $-\log \overline{F}(t)$, all Weibull cumulative hazard functions pass through the point $(1, \lambda)$ regardless of the value of γ.

3.3 Gamma Family

The gamma distribution is another important generalisation of the exponential distribution. The probability density function for the gamma distribution is

$$f(t) = \frac{\lambda^{\gamma} t^{\gamma-1} e^{-\lambda t}}{\Gamma(\gamma)} \ ; \ t > 0, \ \ \lambda > 0, \ \ \gamma > 0$$

where λ and γ are positive parametes denoting scale and shape respectively. Putting $\gamma = 1$ we get the exponential family. It is often difficult to differentiate between Weibull and gamma distributions based on their probability density functions, since shapes of these plots are similar. The differences between these two distributions become apparent when their hazard rates are compared. The behaviour of hazard rate for gamma distribution can only be indirectly investigated as $\overline{F}(t)$ and hence $r(t)$ do not have closed form expressions. The reason that the gamma distribution is less popular in modelling than Weibull is partially attributed to this fact.

$$\begin{aligned} r(t) = \frac{f(t)}{\overline{F}(t)} &= \frac{\lambda^{\gamma} t^{\gamma-1} e^{-\lambda t}}{\Gamma(\gamma)} \left[\int_t^{\infty} \frac{\lambda^{\gamma} x^{\gamma-1} e^{-\lambda x} dx}{\Gamma(\gamma)} \right]^{-1} \\ &= \left[\int_t^{\infty} (\frac{x}{t})^{\gamma-1} e^{-\lambda(x-t)} dx \right]^{-1} \\ &= \left[\int_0^{\infty} \left(\frac{y+t}{t} \right)^{\gamma-1} e^{-\lambda y} dy \right]^{-1} \ \text{for} \ \ 0 < t < \infty, \end{aligned}$$

by putting $x - t = y$.

Thus, $r(t)$ is increasing in t for $\gamma > 1$ and decreasing in t for $0 < \gamma < 1$. Hence the distribution is in the IFR class if $\gamma > 1$ and is in the DFR class if $\gamma < 1$, and of course belongs to both the classes (i.e. the exponential distribution) if $\gamma = 1$. For all values of $\gamma, \lim_{t \to \infty} r(t) = \lambda$, indicating that a lifetime with a gamma distribution will have an exponential tail. Thus, if an item survives far enough into the right-hand tail of the probability density function, the distribution of the remaining time to failure is approximately exponential. The cumulative hazard function and mean residual life function must also be evaluated numerically.

Figure 3.2 shows the hazard curves for gamma distribution with scale parameter equal to one and shape parameter equal to 0.5, 1 and 1.5.

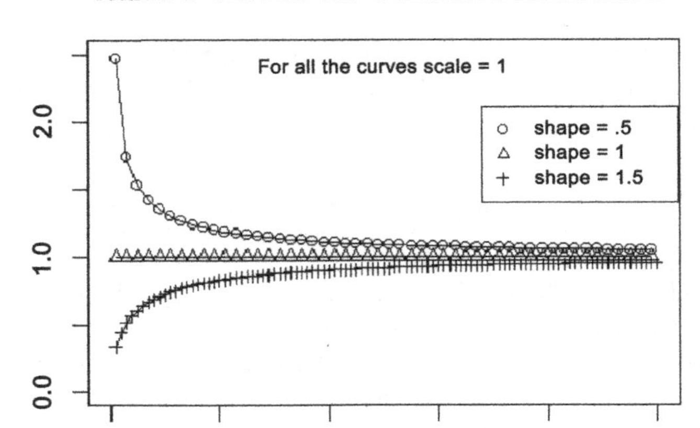

Figure 3.2

The exponential, Weibull and gamma distributions are popular lifetime models. Besides there are several other models which are also useful in modelling lifetime distributions. These are discussed below.

3.4 Log-Normal Family

A random variable T is said to have lognormal distribution when $Y = \log_e T$ is distributed as normal (Gaussian) with mean μ and variance σ^2. The p.d.f. and survival function of lognormal distribution, respectively are:

$$f(t) = \frac{1}{t\sigma\sqrt{2\pi}} \exp\left[-\frac{1}{2\sigma^2}(\log_e t - \mu)^2\right], \quad t > 0, \sigma > 0$$

and

$$\overline{F}(t) = \frac{1}{\sigma\sqrt{2\pi}} \int_t^\infty \frac{1}{x} \exp\left[-\frac{1}{2\sigma^2}(\log_e x - \mu)^2\right] dx.$$

It may be noted that μ and σ^2 which are the location and scale parameters of the normal distribution (of Y) are scale and shape parameters respectively for the lifetime distribution (of T).

The mean and the variance of the distribution are given by

$$E(T) = \exp[\mu + \sigma^2/2]$$

$$Var(T) = [e^{2\mu+\sigma^2}][e^{\sigma^2} - 1].$$

The density curve is positively skew and the skewness increases with σ^2. There is no closed-form expressions for survival and hazard function.

However, computing survival and hazard functions is not difficult. We can write

$$\overline{F}(t) = P\left[Z > \frac{\log_e t - \mu}{\sigma}\right],$$

where $Z \sim N(0,1)$. Thus

$$\overline{F}(t) = 1 - \Phi\left(\frac{\log_e t - \mu}{\sigma}\right)$$

where $\Phi(\cdot)$ represents distribution function of standard normal variate. So using the table of the cumulative probability integral for Z, one can evaluate the survival function of T. Similarly, using the table of ordinates of standard normal distribution we can compute $f(t)$ and thus we can get values for hazard function, $h(t) = \frac{f(t)}{\overline{F}(t)}$.

The hazard function is non-monotonic; initially it increases, reaches a maximum and then decreases to zero as time approaches infinity.

Figure 3.3 represents the plots of hazard function for $\mu = \log(10)$ and (a) $\sigma = 0.4$, (b) $\sigma = 0.6$, (c) $\sigma = 0.8$ and (d) $\sigma = 1$.

This is one of the most widely used probability distributions in describing the life data resulting from a single semiconductor failure mechanism or a closely related group of failure mechanisms. This is a suitable model for patients of tuberculosis or other diseases where the potential for death increases early in the disease and then decreases when the effect of the treatment is evident. Osgood (1958), Feinleib and MacMohan (1960) and Feinleib (1960) observed that the distribution of survival time of several diseases such as Hodgkin's disease and Aronic leukemia too could be rather closely approximated by a log-normal distribution. Horner (1987) showed that the distribution of age at onset of Alzheimer's disease follows lognormal distribution.

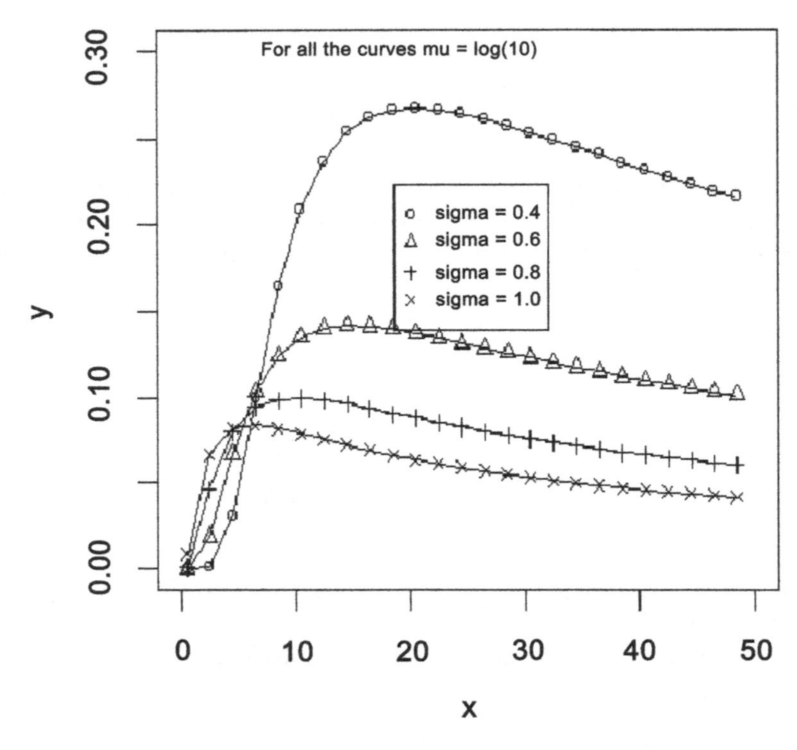

Figure 3.3

By the central limit theorem, the distribution of the product of n independent positive random variables approaches a lognormal distribution under very general conditions. The distribution of the size of an organism whose growth is subjected to many small impulses, the effect of which is proportional to the momentary size of the organism is lognormal by the above result.

The two-parameter lognormal distribution can be generalised to a three-parameter distribution by replacing t with $t - \delta$. In other words, T has three-parameter lognormal distribution if $Y = \log_e(T - \delta)$ follows normal distribution with mean μ and variance σ^2.

3.5 Linear Failure Rate Family

It is given by

$$F(x) = 1 - \exp\{-(x + \frac{1}{2}\theta x^2)\}, x > 0, \theta \geq 0,$$
$$f(x) = (1 + \theta x)e^{-(x + \frac{1}{2}\theta x^2)} \text{ and}$$
$$r(x) = \frac{f(x)}{\overline{F}(x)} = (1 + \theta x).$$

This too is a generalisation of exponential distribution as $\theta = 0$ gives the exponential distribution with failure rate 1. This distribution is a suitable model for items which exhibit positive ageing and has particularly simple formula for the failure rate.

3.6 Makeham Family

This is given by the distribution function

$$F(x) = 1 - \exp[-\{x + \theta(x + e^{-x} - 1)\}], \quad x > 0, \quad \theta \geq 0.$$

$\theta = 0$ again leads to the exponential distribution.

$$f(x) = exp\{-[x + \theta(x + e^{-x} - 1)]\}[1 + \theta(1 - e^{-x})] \text{ and}$$

$$r(x) = [1 + \theta(1 - e^{-x})].$$

It is seen that $r(x)$ is an increasing function of x for $\theta > 0$. Therefore Makeham distribution belongs to IFR class.

3.7 Pareto Family

Simple (one parameter) form of this distribution is given by

$$F(x) = 1 - (1 + \theta x)^{-1/\theta}, \quad x > 0, \theta > 0$$
$$f(x) = (1 + \theta x)^{-(1/\theta + 1)}, \quad x > 0, \quad \theta > 0$$
$$r(x) = (1 + \theta x)^{-1}, \quad x > 0, \quad \theta > 0.$$

It is a family of DFR distributions.

Harris (1968) has pointed out that a two-parameter version of this distribution known as "Pareto distribution of second kind" (sometimes referred as Lomax distribution) arises as a compound exponential distribution when the parameter of the exponential distribution, is itself distributed as a gamma variate. Let

$$P[X \leq x|\theta] = 1 - e^{-x/\theta}, \quad x > 0, \quad \theta > 0$$

and $\mu = 1/\theta$ has a gamma distribution. Then

$$\begin{aligned} F(x) &= P[X \leq x] \\ &= \frac{1}{\beta^{\alpha}\Gamma(\alpha)} \int_0^{\infty} t^{(\alpha-1)} e^{-t/\beta}(1 - e^{-tx})dt \\ &= 1 - (\beta x + 1)^{-\alpha}, \alpha, \beta > 0; x > 0, \\ f(x) &= \beta\alpha(\beta x + 1)^{-(\alpha+1)}, \quad \alpha, \beta > 0; x > 0 \end{aligned}$$

and

$$r(x) = \frac{\alpha\beta}{(\beta x + 1)}, \quad \alpha, \quad \beta > 0, \quad x > 0.$$

Observe that $r(x) \downarrow x$. Hence this distribution also belongs to the DFR class.

Note: All the life distributions considered above have $[0, \infty)$ as their support. But by changing the variable x to $x + \delta$ we can always shift the support to $[\delta, \infty)$. Usually $\delta > 0$ indicates the threshold of lifetime so that the lifetimes smaller than δ are not possible.

3.8 The Distribution of a Specific Parallel System

Consider a two-component parallel system. Suppose that the two components are independent and have respective life distributions:

$$F_1(t) = 1 - e^{-\lambda_1 t} \text{ and } F_2(t) = 1 - e^{-\lambda_2 t}.$$

If F is the life distribution of the system then

$$\overline{F}(t) = 1 - (1 - e^{-\lambda_1 t})(1 - e^{-\lambda_2 t}),$$

so that

$$f(t) = \lambda_1 e^{-\lambda_1 t} + \lambda_2 e^{-\lambda_2 t} - (\lambda_1 + \lambda_2)e^{-(\lambda_1+\lambda_2)t}$$

and

$$r(t) = \frac{\lambda_1 e^{-\lambda_1 t} + \lambda_2 e^{-\lambda_2 t} - (\lambda_1 + \lambda_2) e^{-(\lambda_1+\lambda_2)t}}{e^{-\lambda_1 t} + e^{-\lambda_2 t} - e^{-(\lambda_1+\lambda_2)t}}.$$

It can be verified that $r(t) \uparrow$ on $(0, t_0)$ and decreases on (t_0, ∞) where t_0 depends on λ_1 and λ_2. Figure 3.4 shows this behaviour for various combinations of λ_1 and λ_2, normalised so that $\lambda_1 + \lambda_2 = 1$.

Representative shapes of the failure rate of a system consisting of two exponential components in parallel, with failure rates h_1 and h_2 such that $\lambda_1 + \lambda_2 = 1$.

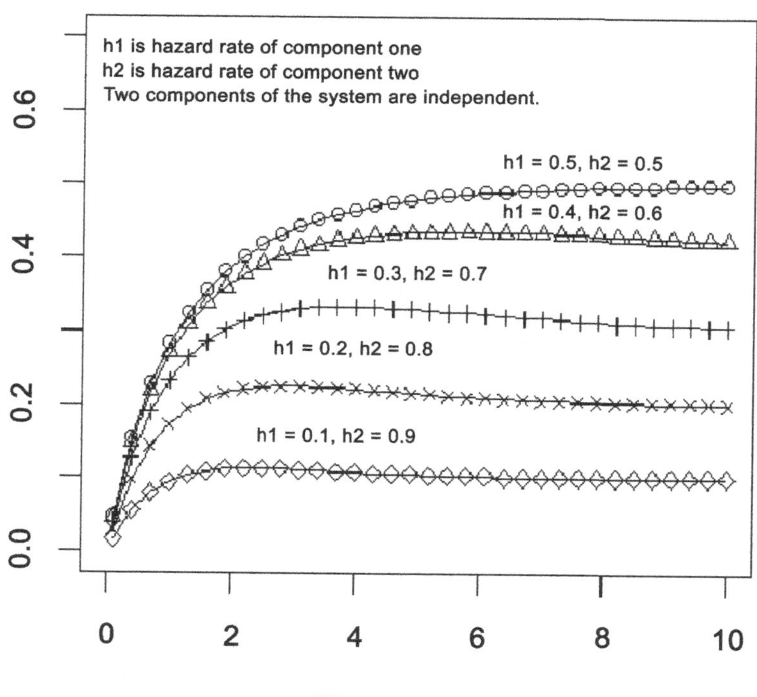

Figure 3.4

3.9 Lehmann Families

This is a very useful family of life distributions generated from a given survival function and extensively used to model the effect of covariates. Let $\overline{F}_0(t)$ be an arbitrary known survival function. If ψ is positive then

$$\overline{S}(t) = [\overline{F}_0(t)]^{\psi}, \quad \psi > 0, \quad t > 0$$

is also a survival function. If, in particular, ψ is the positive integer n, then it represents the survival function of $\min(X_1, ..., X_n)$ where X_i's are i.i.d. r.v.s with $F_0(t)$ as the common distribution function.

The corresponding density is

$$f_{\psi}(t) = \psi[\overline{F}_0(t)]^{\psi-1} f_0(t)$$

and the failure rate is

$$\begin{aligned} r_{\psi}(t) &= \frac{\psi[\overline{F}_0(t)]^{\psi-1} f_0(t)}{[\overline{F}_0(t)]^{\psi}} \\ &= \psi r_0(t) \end{aligned}$$

where $r_0(t)$ is failure rate of F_0. Thus, hazards are proportional. Hence Lehmann family is also known as the proportional hazards family.

3.10 Choice of the Model

The families of distributions outlined above can be judged by

(1) their technical convenience for statistical inference;
(2) the availability of explicit and reasonably simple forms for the survivor, the density and the hazard functions;
(3) the qualitative shape (monotonocity, log concavity, boundedness etc.) of the hazard;
(4) the behaviour of the survival function for small values of time;
(5) the behaviour of the survival function for large times;
(6) any connection with a special stochastic model of failure, etc.

In many applications there will be insufficient information to choose between the different forms by empirical analysis. Then it is legitimate to make the choice on grounds of convenience, if parametric analysis is to be used. Otherwise one can go for non-parametric analysis. Points (1) and (2) are closely related especially when censored data are to be analysed.

Behaviour for small t will be critical for some industrial applications, for instance, where guarantee periods are involved. But in most medical applications the upper tail concerned with relatively long survival times will be of more interest. Properties based on the hazard or the integrated hazard lead directly to methods for analysing censored data. The integrated hazard or log survival function has the advantage of indicating directly the behaviour of the upper tail of the distribution and of leading to a reasonably smooth plot when applied to empirical data.

Guidelines for the choice of the models can also be from the behaviour of $\log T$. For censored data, comparison via the hazard or the loghazard is probably the most widely used approach.

3.11 Some Further Properties of the Exponential Distribution

(a) If $T_1, T_2, ..., T_n$ are independent; $T_i \sim$ exponential (λ_i) for $i = 1, 2, ..., n$ and $T = min\{T_1, T_2, ..., T_n\}$, then

$$T \sim \text{ exponential } (\sum_{i=1}^{n} \lambda_i).$$

First we shall establish a general result which is important in its own right.

If $X_1, X_2, ..., X_n$ are independent r.v.'s then $Z = Min\{X_1, ..., X_n\}$ has failure rate $r(t)$ given by

$$r(t) = \sum_{i=1}^{n} r_i(t)$$

where $r_i(t)$ is failure rate of $X_i (i = 1, 2, ..., n)$.
Proof. Let

$$X_{min} = \min\{X_1, ..., X_n\}.$$

Then,

$$\begin{aligned}
\overline{F}_{X_{min}}(t) &= P[X_{min} > t] \\
&= \prod_{i=1}^{n} P[X_i > t], \quad \text{since } X_i\text{'s are independent} \\
&= \prod_{i=1}^{n} \overline{F}_i(t).
\end{aligned}$$

Hence,

$$\frac{d}{dt}[-\log \overline{F}_{min}(t)] = r(t) = \sum_{i=1}^{n} r_i(t)$$

where $\overline{F}_i(t)$ = survival function X_i.

Specialising the above proof for constant failure rate gives the result (a).

(b) If $T_1, T_2, ..., T_n$ are independent and identically distributed exponential random variables with parameter λ, then

$$2\lambda \sum_{i=1}^{n} T_i \sim \chi^2_{2n}.$$

Proof: Since $T_1, ..., T_n$ are i.i.d.r.v. having exponential distribution with parameter λ,

$$\sum_{i=1}^{n} T_i \sim Gamma(\lambda, n).$$

$$\lambda \sum_{i=1}^{n} T_i \sim Gamma(1, n).$$

$$2\lambda \sum_{i=1}^{n} T_i \sim \chi^2_{2n}.$$

(c) If T is a continuous non-negative r.v. with cumulative failure rate function $H(T)$ then $H(T)$ is exponential with parameter one.

Proof. If $S(t)$ is survival function of T then the survivor function of $H(T)$ is

$$\begin{aligned} P[H(T) > t] &= P[-\log S(T) > t], \quad \text{for } t > 0 \\ &= P[S(T) \leq e^{-t}] \\ &= P[U \leq e^{-t}], \quad \text{for } t > 0, \quad \text{where} \end{aligned}$$

U is uniformly distributed over $(0, 1)$ by the probability integral transform.

Therefore $P[H(T) \geq t] = e^{-t}, t > 0$.

Hence $H(T) \sim$ exponential (1).

(d) If X follows the Weibull distribution with parameters λ and γ then

$Y = X^\gamma$ has the exponential distribution with parameter λ.

$$\begin{aligned} f(x) &= e^{-\lambda x^\gamma}\lambda\gamma x^{\gamma-1}, x \geq 0; \lambda, \gamma > 0. \\ y = x^\gamma &\Rightarrow dy = \gamma x^{\gamma-1}dx. \\ h(y) &= \text{ density of } Y \\ &= \lambda e^{-\lambda y}, y \geq 0, \lambda > 0. \end{aligned}$$

Therefore $Y \sim$ exponential (λ).

(e) Let $T_1, T_2, ..., T_n$ be a random sample from exponential distribution with parameter λ and $T_{(1)}, ..., T_{(n)}$ are corresponding order statistics. Let $Y_i = T_{(i)} - T_{(i-1)}, i = 2, 3, ..., n$ and $Y_1 = T_{(1)}$ be the consecutive sample spacings. Define the normalised sample specings by

$$D_i = (n - i + 1)Y_i, \quad i = 1, 2, ..., n.$$

Then (i) Y_i's are independent exponentially distributed random variables with parameters $(n - i + 1)\lambda, i = 1, 2, ..., n$ respectively.

(ii) D_i's $(i = 1, 2, ..., n)$ are i.i.d. exponential with parameter λ.

Proof. The joint p.d.f. of $T_{(1)}, T_{(2)}, ..., T_{(n)}$ is

$$f_{T_{(1)},...,T_{(n)}}(t_1, t_2, ..., t_n) = n!\lambda e^{-\lambda t_1}\lambda e^{-\lambda t_2}...\lambda e^{-\lambda tn}, 0 < t_1 < t_2... < t_n < \infty.$$

Consider the transformation ϕ from $T_{(1)}, ..., T_{(n)}$ to $Y_1, Y_2, ..., Y_n$:

$$\phi : \begin{cases} Y_1 = T_{(1)} \\ Y_2 = T_{(2)} - T_{(1)} \\ \vdots \\ Y_n = T_{(n)} - T_{(n-1)} \end{cases} \qquad \phi^{-1} : \begin{cases} T_{(1)} = Y_1 \\ T_{(2)} = Y_1 + Y_2 \\ \vdots \\ T_{(n)} = Y_1 + ... + Y_n \end{cases}$$

ϕ is 1-1 transformation from

$$A = \{T_{(1)}, ..., T_{(n)} | 0 < T_{(1)} ... < T_{(n)} < \infty\}$$

to

$$B = \{Y_1, ..., Y_n | Y_i \geq 0, i = 1, 2, ..., n\}$$

with Jacobian of transformation:

$$|J| = \begin{vmatrix} 1\ 0\ 0\ ...\ 0 \\ 1\ 1\ 0\ ...\ 0 \\ 1\ 1\ 1\ ...\ 0 \\ 1\ 1\ 1\ ...\ 1 \end{vmatrix} = 1.$$

So that

$$\begin{aligned} f_{Y_1,\ldots,Y_n}(y_1,\ldots,y_n) &= n!\lambda^n e^{-\lambda y_1} e^{-\lambda(y_1+y_2)} \ldots e^{-\lambda(y_1+y_2+\ldots+y_n)} \\ &= n\lambda e^{-n\lambda y_1}.(n-1)\lambda e^{-(n-1)\lambda y_2} \ldots \lambda e^{-\lambda y_n} \\ &= \prod_{i=1}^{n} f_{Y_i}(y_i), \quad i = 1,2,\ldots,n. \end{aligned}$$

(ii) Y_i's are independent $\Rightarrow$ D_is are independent.
$Y_i \sim Exp((n-i+1)\lambda)$
$D_i \sim (n-i+1)Y_i \sim Exp(\lambda), i = 1,2,\ldots,n.$

(f) If $T_1,\ldots,T_n$ are i.i.d.r.v. with exponential distribution and $T_{(r)}$ is the r-th order statistic, then

$$E(T_{(r)}) = \sum_{k=1}^{r} \frac{1}{(n-k+1)\lambda}$$

and

$$Var[T_{(r)}] = \sum_{k=1}^{r} \frac{1}{[(n-k+1)\lambda]^2}.$$

Note that

$$T_{(r)} = \sum_{i=1}^{r} Y_i$$

where Y_i's are sample spacings. The result now follows by using the means and variances of the Y_i's.

Exercises

Exercise 3.1: Find $h(t)$ and mean time to failure assuming:

$$F(t) = 1 - \frac{8}{7}e^{-t} + \frac{1}{7}e^{-8t} \quad \text{for} \quad t > 0.$$

Exercise 3.2: The failure time of an electronics device is described by a Pearson type V curve. The density function of the failure time is:

$$f(t) = \frac{t^{-(\alpha+1)}e^{-\beta/t}}{\beta^{-\alpha}\Gamma(\alpha)} \quad \text{if } t > 0 \text{ and } f(t) = 0 \text{ otherwise.}$$

Obtain an expression for mean time to failure of the device. What is the mean time to failure if $\alpha = 3$ and $\beta = 4000$ hours.

Exercise 3.3: The failure of a component is lognormally distributed with: $\mu = 6$ and $\sigma = 2$. Find reliability of the component and hazard rate for the life of 200 time units.

Exercise 3.4: A manufacturer uses rotary compressors. Experimental data show that the failure times (between 0 and 1 year) of the compressors follow a beta distribution with $\alpha = 4$ and $\beta = 2$.

What is the mean residual life of a compressor given that the compressor has survived five months?

Exercise 3.5: The density function of Gompertz distribution is:

$$f(t) = \exp\left[(\lambda + \gamma t) - \frac{1}{\gamma}(e^{\lambda+\gamma t} - e^{\lambda})\right] \quad \text{for } t > 0.$$

Find the survival function and hazard function. For what values of γ there is (i) positive ageing (ii) negative ageing and (iii) no ageing?

Chapter 4

Parametric Analysis of Survival Data

4.1 Introduction

In the last chapter we have investigated several parametric distributions which are useful in modelling lifetime data. In this chapter we shall consider the techniques of analysis of lifetime data such as point and interval estimation of the unknown parameters and testing hypotheses regarding these parameters. In general, we shall use the method of maximum likelihood for estimation. The estimators obtained from this method have certain desirable properties and the method is easily applied to censored data as well. We shall first review the basic principles of this technique.

4.2 Method of Maximum Likelihood

Let $T_1, T_2, ..., T_n$ be a random sample from a life distribution having probability density $f(x; \underline{\theta})$ where $\underline{\theta} = (\theta_1, \theta_2, ..., \theta_p) \in \Theta$ is the vector of unknown parameters. Since the lifetimes are independent, the likelihood function $L(\underline{t}, \underline{\theta})$, is the product of probability density functions evaluated at each sample point. Thus,

$$L(\underline{t}, \underline{\theta}) = \prod_{i=1}^{n} f(t_i, \underline{\theta}),$$

where $\underline{t} = (t_1, t_2, ..., t_n)$ is the data point. The maximum likelihood estimator $\hat{\underline{\theta}}$ is the value of $\underline{\theta}$ which maximizes $L(\underline{t}, \underline{\theta})$ for fixed $\underline{t}$. That is, $\hat{\underline{\theta}}$ is the maximum likelihood estimator of $\underline{\theta}$, if $L(t, \hat{\underline{\theta}}) \geq L(t, \tilde{\underline{\theta}})$ for any other estimator or value $\tilde{\underline{\theta}}$ of $\underline{\theta}$. One may say that $f_{\hat{\underline{\theta}}}(t)$ corresponds to the distribution that is most likely to have produced the data $t_1, t_2, ..., t_n$ in the family $\{f_{\underline{\theta}}, \underline{\theta} \in \Theta\}$.

In practice, it is often easier to maximize the log likelihood function $\log L(\underline{t}, \underline{\theta}) = \sum_{i=1}^{n} \log f(t_i, \underline{\theta})$ to find the vector of maximum likelihood estimators. It is a valid procedure because the logarithm function is monotonically increasing. There is an added advantage that $\log L(\underline{t}, \underline{\theta})$ as a function of $\underline{t}$ is asymptotically normally distributed by the central limit theorem, being the sum of n independent identically distributed random terms, under well-known regularity conditions.

Since $L(\underline{t}, \underline{\theta})$ is a joint density function, it must integrate over the range of $\underline{t}$ to 1. Therefore,

$$\int_0^\infty \int_0^\infty \dots \int_0^\infty L(\underline{t}, \underline{\theta}) d\underline{t} = 1. \tag{4.2.1}$$

Under regularity conditions which allow interchange of differentiation and integration operations, the partial derivative of the left side with respect to one of the parameters, θ_i, yields

$$\frac{\delta}{\delta\theta_i} \int_0^\infty \dots \int_0^\infty L(\underline{t}, \underline{\theta}) d\underline{t} = \int_0^\infty \dots \int_0^\infty \frac{\delta}{\delta\theta_i} \log L(\underline{t}, \underline{\theta}) L(\underline{t}, \underline{\theta}) \underline{dt}$$

$$= E\left[\frac{\delta}{\delta\theta_i} \log L(\underline{t}, \underline{\theta})\right] = E[U_i(\underline{\theta})], \quad i = 1, 2, \dots, p, \tag{4.2.2}$$

where $U(\underline{\theta}) = (U_1(\underline{\theta}), \dots, U_p(\underline{\theta}))'$ is often called the <u>score vector</u>. The argument $\underline{t}$ is suppressed for compactness. Differentiating the right side of (4.2.1) with respect to $\underline{\theta}$ and using (4.2.2) we get,

$$E(U_i(\underline{\theta})) = 0, \quad i = 1, 2, \dots, p \tag{4.2.3}$$

or in the vector form $E[U(\underline{\theta})] = \underline{0}$.

Further differentiation of (4.2.2) with respect to θ_j yields

$$E[U_i(\underline{\theta}) U_j(\underline{\theta})] = E\left[\frac{-\delta^2 \log L(\underline{t}, \underline{\theta})}{\delta\theta_i \delta\theta_j}\right] \quad \begin{matrix} i = 1, 2, \dots, p, \\ j = 1, 2, \dots, p. \end{matrix} \tag{4.2.4}$$

From (4.2.3) and (4.2.4) it follows that

$$E\left[\frac{-\delta^2 \log l(\underline{t}, \underline{\theta})}{\delta\theta_i \delta\theta_j}\right] = cov(U_i(\underline{\theta}), U_j(\underline{\theta})), \begin{matrix} i = 1, 2, \dots, p \\ j = 1, 2, \dots, p \end{matrix}.$$

These elements form the $p \times p$ Fisher information matrix, $I(\underline{\theta})$, whose diagonal elements are the variances and the off-diagonal elements are the covariances of the score vector.

The solutions of the simultaneous likelihood equations,

$$U_i(\underline{\theta}) = \frac{\delta}{\delta\theta_i} \log L(\underline{t}, \underline{\theta}) = 0,$$

are $\hat{\theta}_i$, the maximum likelihood estimators of $\theta_i, i = 1, 2, ..., p$.

The estimators $\hat{\theta}_1, ..., \hat{\theta}_p$, under certain regularity conditions are asymptotically normally distributed with mean $\theta_1, ..., \theta_p$ and variance covariance matrix given by

$$V(\hat{\underline{\theta}}) = \{I(\underline{\theta})\}^{-1}.$$

The observed (sample) information matrix called $i(\underline{\theta})$ is defined by the elements

$$\left(-\frac{\delta^2}{\delta\theta_i \delta\theta_j} \log L(\underline{t}, \underline{\theta}), i, j = 1, ..., p\right).$$

So that $E[i(\underline{\theta})] = I(\underline{\theta})$ at $\underline{\theta} = \hat{\underline{\theta}}$.

Three broad types of asymptotic procedures, based on the likelihood function, are available for testing of hypothesis $\underline{\theta} = \underline{\theta}_0$.

(i) *Wilks Likelihood Ratio*

Let $L(\hat{\underline{\theta}}) = [L(\underline{\theta})]_{\underline{\theta}=\hat{\underline{\theta}}}$.

$$-2 \log \frac{L(\underline{\theta}_0)}{L(\hat{\underline{\theta}})} \xrightarrow{a} \chi_p^2 \text{ under } H_0,$$

where $\xrightarrow{a}$ denotes "asymptotically distributed as".

(ii) *Wald's method based on MLE's*

$$(\hat{\underline{\theta}} - \underline{\theta}_0)' I(\underline{\theta}_0)(\hat{\underline{\theta}} - \underline{\theta}_0) \xrightarrow{a} \chi_p^2 \text{ under } H_0.$$

(iii) *Rao's Scores method*

$$\left[\frac{\delta}{\delta\underline{\theta}} \log L(\underline{\theta}_0)\right]' I^{-1}(\underline{\theta}_0) \left[\frac{\delta}{\delta\underline{\theta}} \log L(\underline{\theta}_0)\right] \xrightarrow{a} \chi_p^2$$

under H_0.

Notice that Rao's method does not use the MLE and hence is recommended in practice if interest is only in hypothesis testing. However, in addition to tests, we usually want estimates and confidence intervals, so we need to compute $\hat{\underline{\theta}}$ anyway. Once we have $\hat{\underline{\theta}}$ and $I(\underline{\theta}_0)$, the Wald method is easy.

The three procedures are asymptotically equivalent and will often give virtually identical conclusions. This is evidenced by the following figure (Figure 4.1) in which we represent the likelihood function of a single parameter for the observed data $\underline{t}$.

Likelihood Function of a single parameter

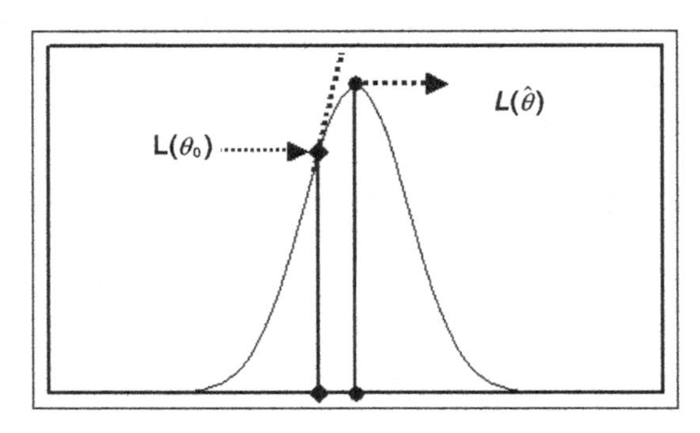

Figure 4.1

The likelihood ratio approach compares the values of the likelihood function at $\hat{\theta}$ and at θ_0, the value provided by the null hypothesis. The Wald approach directly compares $\hat{\theta}$ with θ_0 and the scores approach compares the slope of the likelihood function at θ_0 with the slope at $\hat{\theta}$ (which is zero). Hence nearness or otherwise of $\hat{\theta}$ (the best value of θ that the data provides) from θ_0 (the value of θ under the null hypothesis) can be judged in any of the above ways.

In the above testing procedures we have assumed that the null hypothesis is simple. Tests for composite null hypotheses can also be based on either the score statistic, the Wald statistic or the likelihood ratio statistic. The most common composite hypotheses are those given by simple hypotheses about certain components of $\underline{\theta}$ while leaving the other parameters unspecified as nuisance parameters. We will outline only these methods here.

Suppose that $\underline{\theta}' = (\underline{\theta}_1', \underline{\theta}_2')'$ where $\underline{\theta}_1$ is any $r \times 1$ vector. $1 \leq r < p$ and we wish to test $H_0 : \underline{\theta}_1 = \underline{\theta}_{10}$. If $U'(\underline{\theta}) = \{U_1'(\underline{\theta}), U_2'(\underline{\theta})\}'$ with the components of U_1 given by $U_{1j}(\underline{\theta}) = \frac{\delta}{\delta\theta_j} L(\underline{\theta}), 1 \leq j \leq r$; a test can be based on the vector $U_1((\underline{\theta}_{10}', \overline{\underline{\theta}}_2')')$; where $\overline{\underline{\theta}}_2$ is the restricted maximum likelihood estimator of $\underline{\theta}_2$ computed under the constraint $\underline{\theta}_1 = \underline{\theta}_{10}$.

The conditional distribution of $U_1((\underline{\theta}'_{10}, \overline{\underline{\theta}}'_2)')$ given $\underline{\theta}_2 = \overline{\underline{\theta}}_2$ can be used for critical values for the test statistic and, when $\overline{\underline{\theta}}_2$ is unique, this will be the same as the conditional distribution of $U_1((\underline{\theta}'_{10}, \overline{\underline{\theta}}'_2)')$ given $U_2 = \underline{0}$. Since $U(\underline{\theta})$ is asymptotically multivariate normal, this conditional distribution will be the r-variate normal distribution with mean zero and covariance matrix $\sum_{10}\{(\underline{\theta}'_{10}, \overline{\underline{\theta}}_2)'\} = I_{11} - I_{12}I_{22}^{-1}I_{21}$ where

$$I(\underline{\theta}) = \begin{pmatrix} I_{11}(\underline{\theta}), \ I_{12}(\underline{\theta}) \\ I_{21}(\underline{\theta}), \ I_{22}(\underline{\theta}) \end{pmatrix}$$

(where if $I_{22}(\underline{\theta})$ is singular, $\{I_{22}(\underline{\theta}\}^{-1}$ should be replaced by a generalised inverse).

Tests for $\underline{\theta}_1 = \underline{\theta}_{10}$ can then be based on the three statistics as follows:

(i) the likelihood ratio statistic Λ is given by

$$\Lambda = \frac{\max_{\underline{\theta} \in \Theta_0} L(\underline{\theta})}{\max_{\underline{\theta} \in \Theta} L(\underline{\theta})},$$

where

$$\Theta_0 = \{(\underline{\theta}'_1, \underline{\theta}'_2)'; \underline{\theta}'_1 = (\theta_{10}, ..., \theta_{r0})\}$$

and $-2\log\Lambda$ has asymptotically χ^2 distribution with r degrees of freedom.

(ii) Wald statistic given by $(\hat{\underline{\theta}}_1 - \underline{\theta}_{10})' \sum_{10}(\hat{\underline{\theta}}_1 - \underline{\theta}_{10})$ also has χ^2 distribution with r degrees of freedom and

(iii) $U'_1\{(\underline{\theta}'_{10}, \overline{\underline{\theta}}'_2)\} \sum_{10}^{-1} U_1\{(\underline{\theta}'_{10}, \overline{\theta}'_2)'\}$ will have an approximate χ^2 distribution with r degrees of freedom.

Iterative procedures for solving a system of likelihood equations

The following are the two commonly used methods for obtaining MLEs when closed form solutions are not possible.

(i) *Newton - Raphson Method*: Assume $\hat{\underline{\theta}}^{(0)} = (\hat{\theta}_1^{(0)}, ..., \hat{\theta}_p^{(0)})'$ is an initial guess at the solution. Then

$$\hat{\underline{\theta}}^{(1)} = \hat{\underline{\theta}}^{(0)} + (i(\underline{\theta}^{(0)}))^{-1}\frac{\delta}{\delta\theta}\log L(\underline{\theta}^{(0)}),$$

where

$$i(\underline{\theta}^{(0)}) = [i(\underline{\theta})]_{\underline{\theta}=\underline{\theta}^{(0)}}$$

and

$$L(\underline{\theta}^{(0)}) = L[(\underline{\theta})]_{\underline{\theta}=\underline{\theta}^{(0)}}.$$

In general,

$$\hat{\underline{\theta}}^{(j+1)} = \hat{\underline{\theta}}^{(j)} + (i(\underline{\theta}^{(j)})^{-1}\frac{\delta}{\delta\underline{\theta}}\log L(\underline{\theta}^{(j)}), \quad j = 1, 2, ... \tag{4.2.5}$$

(ii) *Fisher's Method of Scoring*: Replacing sample information matrix $i(\underline{\theta})$ in (4.2.5) by Fisher's information matrix we get the following iterative formula for Fisher's method:

$$\hat{\underline{\theta}}^{(j+1)} = \hat{\underline{\theta}}^{(j)} + (I(\underline{\theta})^{(j)})^{-1}\frac{\delta}{\delta\underline{\theta}}\log L(\underline{\theta}^{(j)}), \quad j = 1, 2, ... \tag{4.2.6}$$

Fisher's method of scoring produces improved convergence in some instances. However, in many situations, particularly if censoring is present, $I(\underline{\theta})$ is not mathematically tractable. Hence the Newton-Raphson method is used.

4.3 Parametric Analysis for Complete Data

In what follows we shall discuss the parametric analysis for complete data.
(A) The Exponential Distribution

Let $t_1, t_2, ..., t_n$ be a random sample from an exponential distribution with parameter λ.

$$f(t; \lambda) = \lambda e^{-\lambda t}, t \geq 0; \lambda > 0.$$

$$L(\underline{t}; \lambda) = \prod_{i=1}^{n} \lambda e^{-\lambda t_i} = \lambda^n e^{-\lambda \sum_{i=1}^{n} t_i}.$$

The log likelihood function is

$$\log L(\underline{t}, \lambda) = n \log \lambda - \lambda \sum_{i=1}^{n} t_i.$$

The score is

$$U(\lambda) = \frac{\delta}{\delta\lambda}\log L(\underline{t}, \lambda)$$
$$= \frac{n}{\lambda} - \sum_{1}^{n} t_i.$$

$$\left[\frac{\delta}{\delta\lambda}\log L(\underline{t},\lambda)\right]_{\lambda=\hat{\lambda}} = 0 \Rightarrow \hat{\lambda} = \frac{n}{\sum_{1}^{n} t_i}.$$

$$I(\lambda) = \frac{n}{\lambda^2}.$$

Sample information at $\hat{\lambda}$ is $\frac{n}{\hat{\lambda}^2}$ and $var(\hat{\lambda})$ is $\frac{\hat{\lambda}^2}{n}$.

Notice that the maximum likelihood estimator of λ is the ratio of the total number of failures to the total lifetime of all the units, i.e. the total time on test. If μ is the mean of the distribution then its maximum likelihood estimator (MLE) is $1/\hat{\lambda}$ which is also the method of moments estimator of μ. It is seen that $\sum_{1}^{n} T_i$ is minimal sufficient statistic. $\overline{T}$ is consistent for μ and $\frac{1}{\overline{T}}$ is a consistent estimator of λ. The asymptotic distribution of $\hat{\lambda}$ is normal with mean λ and variance $\frac{\lambda^2}{n}$. So that

$$\frac{\sqrt{n}(\hat{\lambda}-\lambda)}{\lambda} \xrightarrow{a} N(0,1). \tag{4.3.1}$$

The exact distribution of $\hat{\mu} = \frac{1}{\hat{\lambda}}$ can be derived using the following result: $\sum_{1}^{n} T_i$ is the sum of n independent exponential random variables, hence it has gamma distribution and therefore $\frac{2n\overline{T}}{\mu} = \frac{2n\hat{\mu}}{\mu}$ has χ^2_{2n} distribution. Equivalently $\frac{2n\lambda}{\hat{\lambda}}$ has χ^2_{2n} distribution. From the above result we have

$$E\left[\frac{2n\hat{\mu}}{\mu}\right] = 2n \Rightarrow E(\hat{\mu}) = \mu.$$

Exact Confidence Interval for λ is obtained by using the pivotal quantity $\frac{2n\lambda}{\hat{\lambda}}$. Let $(1-\alpha)$ be the confidence coefficient and $\chi^2_{\alpha/2,2n}$ and $\chi^2_{1-\alpha/2,2n}$ be such that

$$P[\chi^2_{2n} \le \chi^2_{\alpha/2,2n}] = P[\chi^2_{2n} \ge \chi^2_{1-\alpha/2,2n}] = \alpha/2.$$

Then $100(1-\alpha)\%$ equal tailed confidence interval for λ is obtained from:

$$(\chi^2_{\alpha/2,2n} \le \frac{2n\lambda}{\hat{\lambda}} \le \chi^2_{1-\alpha/2,2n}) = 1-\alpha.$$

The required confidence interval (C.I.) is

$$(\frac{\hat{\lambda}}{2n}\chi^2_{\alpha/2,2n}; \frac{\hat{\lambda}}{2n}\chi^2_{1-\alpha/2,2n}).$$

Large Sample Confidence Intervals

(a) From likelihood ratio statistic we have

$$2[\log L(\hat{\lambda}) - \log L(\lambda)] \stackrel{a}{\rightarrow} \chi^2_1.$$

So the required C.I. is obtained by solving the equation

$$2[\log L(\hat{\lambda}) - \log L(\lambda)] = \chi^2_{(1-\alpha)}.$$

(b) From the asymptotic normality of $\hat{\lambda}$;

$$\frac{\hat{\lambda} - \lambda}{\sqrt{\frac{1}{I(\lambda)}}} \stackrel{a}{\rightarrow} N(0,1). \qquad (*)$$

This gives the $100(1-\alpha)\%$ C.I. as

$$\left(\frac{\sqrt{n}\hat{\lambda}}{\sqrt{n} + z_{1-\alpha/2}}, \frac{\sqrt{n}\hat{\lambda}}{\sqrt{n} - z_{1-\alpha/2}}\right)$$

where $z_{1-\alpha/2}$ is such that $Z \le z_{1-\alpha/2} = 1 - \alpha/2$ where Z is the standard normal variable. For,

$$P\left[-z_{1-\alpha/2} \le \frac{\sqrt{n}(\hat{\lambda} - \lambda)}{\lambda} \le z_{1-\alpha/2}\right] = 1 - \alpha.$$

$$P\left[1 - \frac{z_{1-\alpha/2}}{\sqrt{n}} \le \frac{\hat{\lambda}}{\lambda} \le 1 + \frac{z_{1-\alpha/2}}{\sqrt{n}}\right] = 1 - \alpha.$$

$$P\left[\frac{\sqrt{n}\hat{\lambda}}{\sqrt{n} + z_{1-\alpha/2}} \le \lambda \le \frac{\sqrt{n}\hat{\lambda}}{\sqrt{n} - z_{1-\alpha/2}}\right] = 1 - \alpha.$$

However, if $I(\lambda)$ is replaced by $i(\hat{\lambda})$, its consistent estimator, we get from (*),

$$(\hat{\lambda} - z_{1-\alpha/2}\frac{\hat{\lambda}}{\sqrt{n}}, \hat{\lambda} + z_{1-\alpha/2}\frac{\hat{\lambda}}{\sqrt{n}})$$

as $100(1-\alpha)\%$ confidence interval for λ.

Illustration 4.1: Data on Earthquakes (Hand, et al. (1993)).

The following are the time in days between successive serious earthquakes worldwide. An earthquake is included in the data set if its magnitude was at least 7.5 on Richter scale, or if over 1000 people were killed. Recording starts on 16-th of December 1902 and ends on 14-th March 1997. There were 63 earthquakes recorded altogether, and so 62 waiting times.

840, 157, 145, 44, 33, 121, 150, 280, 434, 736, 584, 887, 263, 1901, 695, 294, 562, 721, 76, 710, 46, 402, 194, 759, 319, 460, 40, 1336, 335, 1334, 454, 36, 667, 40, 556, 99, 304, 375, 567, 139, 780, 203, 436, 30, 384, 129, 9, 209, 599, 83, 832, 328, 246, 1617, 638, 937, 735, 38, 365, 92, 82, 220.
Source: The Open University (1981) S237: The earth: Structure, Composition, and Evaluation.

Assume that the earthquakes occur at random and hence waiting times are exponentially distributed. Obtain

- Point estimate of scale parameter (λ),
- Interval estimate of (λ) for confidence coefficient of 95%,

Check the assumption of exponentiality using simple graphical methods.
Solution: $\hat{\lambda} - 0.002289, LCL = 0.001754, UCL = 0.002719$.
Graphical Methods for Checking Exponentiality

1. We plot empirical and estimated survival curves on the same graph paper. If the two curves are close then the model is appropriate.

2. We plot -log(s(t)), where s(t) is empirical survival function, versus t. If data are from exponential distribution, the graph will show linear trend.

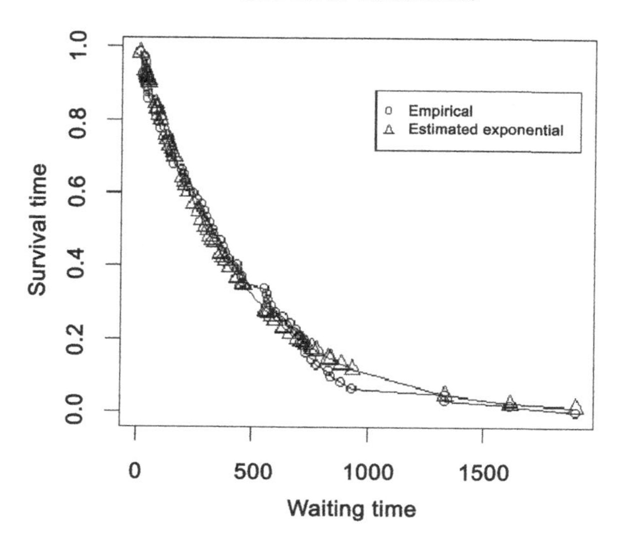

Figure 4.2

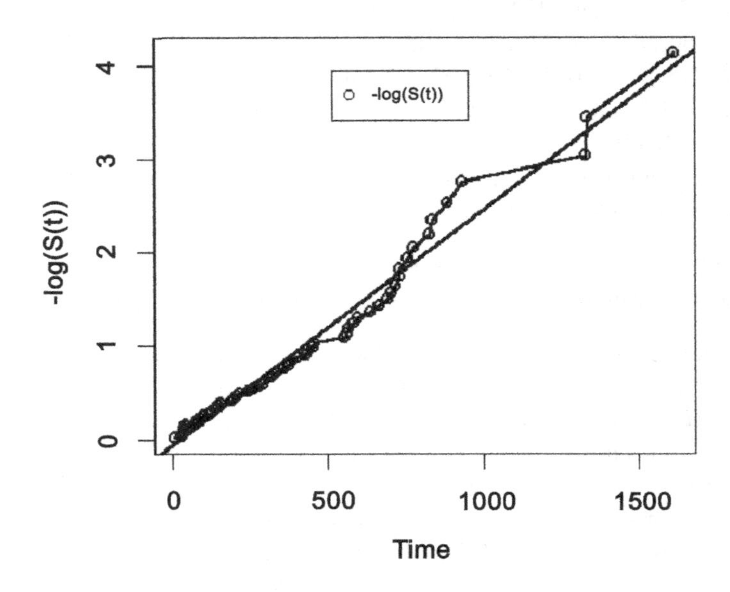

Figure 4.3

The fit of exponential distribution seems to be good.

(B) The Gamma Distribution

Let $t_1, t_2, ..., t_n$ be a random sample from a gamma distribution with scale parameter λ and shape parameter γ

$$f(t; \lambda, \gamma) = \frac{\lambda^\gamma}{\Gamma(\gamma)} e^{-\lambda t} (t)^{\gamma-1}; \quad t \geq 0; \lambda, \gamma > 0.$$

$$L(\underline{t}; \lambda, \gamma) = \frac{\lambda^{n\gamma}}{[\Gamma(\gamma)]^n} e^{-\lambda \sum\limits_1^n t_i} \prod_{i=1}^{n} t_i^{\gamma-1}.$$

$$\log L(\underline{t}; \lambda, \gamma) = n\gamma \log \lambda - n \log \Gamma(\gamma) - \lambda \sum_{i=1}^{n} t_i + (\gamma - 1) \sum_{i=1}^{n} \log t_i.$$

The score vector has components;

(i) $\frac{\delta}{\delta\lambda} \log L(\lambda, \gamma) = \frac{n\gamma}{\lambda} - \sum\limits_1^n t_i$ and

(ii) $\frac{\delta}{\delta\gamma} \log L(\lambda, \gamma) = n \log \lambda - \frac{n\Gamma'(\gamma)}{\Gamma(\gamma)} + \sum\limits_1^n \log t_i$, where $\Gamma'(\gamma) = \frac{\delta}{\delta\gamma}\Gamma(\gamma)$.

The MLEs of λ and γ satisfy

$$\hat{\lambda} = \hat{\gamma}(\bar{t})^{-1} \tag{4.3.2}$$

and

$$n \log \hat{\lambda} + \sum_{1}^{n} \log t_i = \frac{n\Gamma'(\gamma)}{\Gamma(\gamma)}. \tag{4.3.3}$$

Substituting for $\hat{\lambda}$ in (4.3.3) from (4.3.2), we get

$$n \log(\frac{\hat{\gamma}}{\bar{t}}) + \sum_{1}^{n} \log t_i = \frac{n\Gamma'(\hat{\gamma})}{\Gamma(\hat{\gamma})}. \tag{4.3.4}$$

Or

$$\frac{\Gamma'(\hat{\gamma})}{\Gamma(\hat{\gamma})} - \log(\hat{\gamma}) = \log R, \tag{4.3.5}$$

where

$$R = \frac{(\prod_1^n t_i)^{1/n}}{\bar{t}}$$
$$= \text{Ratio of the geometric mean and the arithmetic mean.}$$

Some iterative numerical method such as Newton-Raphson procedure must be used for solving equation (4.3.5). However, Wilk, Gnanadesikan and Huyett (1962) have provided tables for the values of $\hat{\gamma}$ against given values of $(1-R)^{-1}$. These tables are reproduced in Gross and Clark (1975), Bain and Englehardt (1991) and in Deshpande, Gore and Shanubhogue (1995). Thus, solution of (4.3.5) is obtained by using the tables. It may be noted that the intermediate value of $(1 - R)^{-1}$, not available in the table, can be obtained by linear interpolation. After $\hat{\gamma}$ is determined, $\hat{\lambda}$ is obtained from (4.3.2).

Remark. The gamma distribution is a member of the exponential family. The arithmetic mean and the geometric mean form a set of complete sufficient statistics for (λ, γ). Hence MLEs are functions of these statistics.

As usual, MLEs are somewhat biased for small n but become nearly unbiased and efficient for large n. Of course, the question of bias depends on what parameters or functions of parameters are of interest. For example, the MLE of the mean $(= \gamma/\lambda)$ is $\hat{\gamma}/\hat{\lambda}$ where $\hat{\gamma}/\hat{\lambda}$ is sample mean. It is known that sample mean is unbiased estimator of population mean.

Sample information matrix

$$\frac{\delta^2}{\delta\lambda^2}\log L(\lambda,\gamma) = -\frac{n\gamma}{\lambda^2}.$$

$$\frac{\delta^2}{\delta\lambda\delta\gamma}\log L(\lambda,\gamma) = \frac{n}{\lambda}.$$

$$\frac{\delta^2}{\delta\gamma^2}\log L(\lambda,\gamma) = -n\left[\frac{\Gamma''(\gamma)}{[\Gamma(\gamma)]} - \frac{[\Gamma'(\gamma)]^2}{[\Gamma(\gamma)]^2}\right].$$

$$i(\lambda,\gamma) = \begin{bmatrix} \frac{n\gamma}{\lambda^2} & -\frac{n}{\lambda} \\ -\frac{n}{\lambda} & n[\frac{\Gamma''(\gamma)}{(\Gamma(\gamma))} - \frac{[\Gamma'(\gamma)]^2}{\Gamma(\gamma)^2}] \end{bmatrix}.$$

Illustration 4.2: (Birnbaum and Saunders (1958))

In the study of lifetime distribution of aluminium coupon, 17 sets of six strips were placed in specially designed machine. Periodic loading was applied to the strips with a frequency of 18 cycles per second and maximum stress of 21,000 psi. The 102 strips were run till all of them failed. One of the 102 strips tested had to be discarded for an extraneous reason, yielding 101 observations. The data are given below:

370	1055	1270	1502	1763
706	1085	1290	1505	1768
716	1102	1293	1513	1781
746	1102	1300	1522	1782
785	1108	1310	1522	1792
797	1115	1313	1530	1820
844	1120	1315	1540	1868
855	1134	1330	1560	1881
858	1140	1355	1567	1890
886	1199	1390	1578	1893
886	1200	1416	1594	1895
930	1200	1419	1602	1910
960	1203	1420	1604	1923
988	1222	1420	1608	1940
990	1235	1450	1630	1945
1000	1238	1452	1642	2023
1010	1252	1475	1674	2100
1016	1258	1478	1730	2130

1018	1262	1481	1750	2215
1020	1269	1485	1750	2268
				2440

Assume the data are from gamma distribution and estimate the two parameters of the distribution.
Solution:

Arithmetic mean $= A = 1400.911$, Geometric mean $= G = 1342.259$

$$(1 - R)^{-1} = 23.88532.$$

Using Wilk, Gnanadesikan and Huyett tables and interpolating linearly between the values corresponding to $(1 - R)^{-1} = 20$ and $(1 - R)^{-1} = 30$ we get $\hat{\gamma} = 11.85461$. Then using (4.3.2) we get $\hat{\lambda} = 0.008462$.
Note: R-commands for the solution are given in the Appendix of this chapter. Figure 4.4 shows estimated and empirical distribution functions. R-commands for plotting the figure are also given in the Appendix. We can see from the plot that the fit of gamma distribution to the data is good.

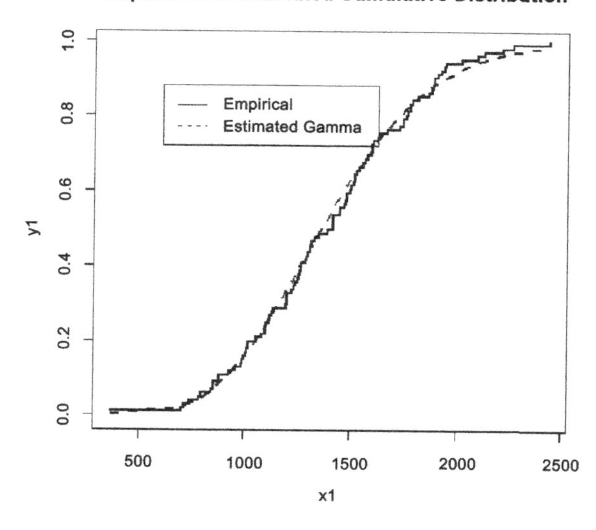

Figure 4.4

(C) The Weibull Distribution

Let $t_1, t_2, ..., t_n$ be a random sample from the Weibull distribution with scale parameter λ and shape parameter γ.

$$f(t; \lambda, \gamma) = \lambda\gamma t^{\gamma-1}\exp(-\lambda t^{\gamma}), \quad t \geq 0, \quad \lambda, \quad \gamma > 0$$

$$\log L(\lambda, \gamma) = n\log\lambda + n\log\gamma + \sum_{i=1}^{n}(\gamma - 1)\log t_i - \lambda\sum_{1}^{n} t_i^{\gamma}.$$

The elements of the score vector are

(i) $\frac{\delta}{\delta\lambda}\log L(\underline{t}; \lambda, \gamma) = \frac{n}{\lambda} - \sum\limits_{1}^{n} t_i^{\gamma}$ and

(ii) $\frac{\delta}{\delta\gamma}\log L(\underline{t}; \lambda, \gamma) = \frac{n}{\gamma} + \sum\limits_{1}^{n}\log t_i - \lambda\sum\limits_{1}^{n} t_i^{\gamma}\log t_i$.

The MLEs of λ and γ satisfy the equations:

$$\frac{n}{\hat{\lambda}} - \sum_{1}^{n} t_i^{\hat{\gamma}} = 0. \tag{4.3.6}$$

$$\frac{n}{\hat{\gamma}} + \sum_{1}^{n}\log t_i - \hat{\lambda}\sum_{1}^{n} t_i^{\hat{\gamma}}\log t_i = 0. \tag{4.3.7}$$

From (4.3.6),

$$\hat{\lambda} = n[\sum_{1}^{n} t_i^{\hat{\gamma}}]^{-1} \tag{4.3.8}$$

and from (4.3.7)

$$\hat{\lambda} = [\frac{n}{\hat{\gamma}} + \sum_{1}^{n}\log t_i][\sum_{1}^{n} t_i^{\hat{\gamma}}\log t_i]^{-1}. \tag{4.3.9}$$

Thus

$$n[\sum_{1}^{n} t_i^{\hat{\gamma}}]^{-1} = [\frac{n}{\hat{\gamma}} + \sum_{1}^{n}\log t_i][\sum_{1}^{n} t_i^{\hat{\gamma}}\log t_i]^{-1}$$

or

$$n[\sum_{1}^{n} t_i^{\hat{\gamma}}]^{-1}[\sum_{1}^{n} t_i^{\hat{\gamma}}\log t_i] - \frac{n}{\hat{\gamma}} - \sum_{1}^{n}\log t_i = 0.$$

That is

$$h(\hat{\gamma}) = 0. \tag{4.3.10}$$

Solution of (4.3.10) is obtained by using Newton-Raphson or similar numerical method and then $\hat{\lambda}$ is obtained by substituting the value of $\hat{\gamma}$, thus obtained, in (4.3.8). However, in order to solve (4.3.10) by numerical methods, an initial or starting solution is required. The following graphical method may be used to get a starting solution.

Graphical Procedure for Estimating the Parameters:

The survival function of the Weibull distribution is

$$\overline{F}(t) = \exp(-\lambda t^{\gamma}).$$

Hence

$$\log\{[\overline{F}(t)]^{-1}\} = \lambda t^{\gamma}$$

and therefore

$$\log\log[\{\overline{F}(t)\}^{-1}] = \log\lambda + \gamma\log t.$$

Let $t_{(1)} < t_{(2)} ... < t_{(n)}$ be the order statistics from the random sample. Estimate $\overline{F}(t_{(i)})$ by $\hat{\overline{F}}(t_{(i)})$ where $\hat{\overline{F}}(t_{(i)})$ is empirical survival function. Plot $\log\log[\{\hat{\overline{F}}(t_{(i)})\}^{-1}]$ against $\log[t_{(i)}]$ for $i = 1, 2, ..., n$. If the underlying distribution is indeed Weibull, the graph will be approximately a straight line. A line could be fitted by the usual least squares techniques or just by inspection. The slope of the line will give the initial estimate of γ and the y-intercept will provide an initial estimate of λ.

Sample Information Matrix

$$i(\lambda,\gamma) = \begin{bmatrix} \frac{n}{\lambda^2} & \sum_1^n t_i^{\gamma}\log t_i \\ & \frac{n}{\gamma^2} + \lambda\sum_1^n t_i^{\gamma}(\log t_i)^2 \end{bmatrix}.$$

Alternatively, $(\hat{\lambda},\hat{\gamma})$ can be obtained by using Newton-Raphson method of scoring with the estimates obtained from the graphical method as the initial solution. Let $\underline{\theta} = (\lambda,\gamma)'$ and $\underline{\theta}^{(0)} = (\hat{\lambda},\hat{\gamma})'$ as obtained from graphical method. The iterative procedure is as given in (4.2.5).

Illustration 4.3

The following are the times (in minutes) taken for an insulating fluid between electrodes recorded at voltage 36 kV. Assume Weibull distribution and estimate the parameters of the distribution.

.35, .59, .96, .99, 1.69, 1.97, 2.07, 2.58, 2.71, 2.90, 3.67, 3.99, 5.35, 13.77

Solution:

Graphical Method.

Figure 4.5 shows the graph of $\log[-\log F_n(t(i))]$ vs. $\log[t(i)]$.

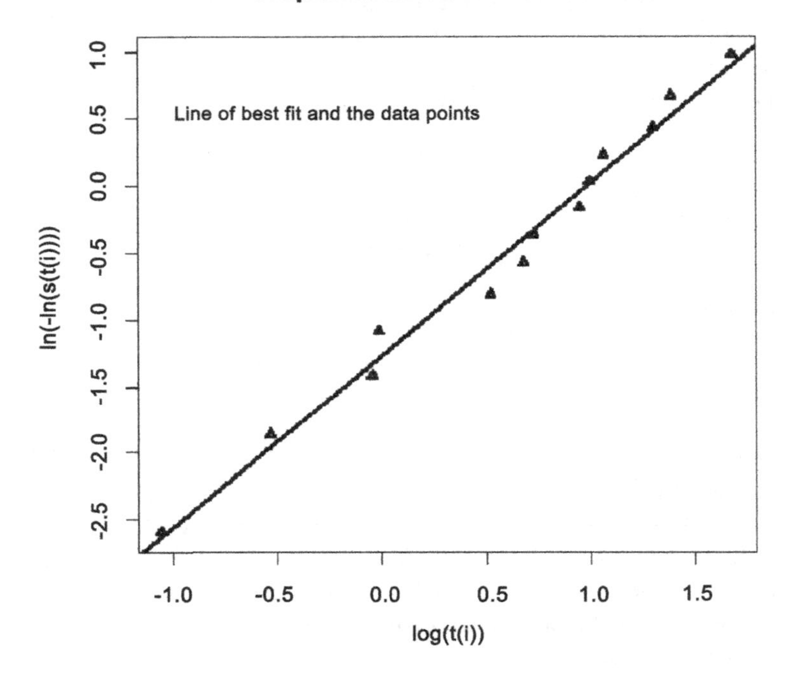

Figure 4.5

The points show strong linear trend. A line of best fit, obtained by the method of least squares, is also shown on the plot.

Correlation coefficient = 0.9847

Regresion coefficient = 1.29732.

The assumption of Weibull model is justifiable. From the graph, estimate of $\gamma = 1.29732$. In order to get refined estimate of γ we use approximate trial and error method. For this we compute $h(\hat{\gamma})$ for values of $\hat{\gamma}$ in the range $(1, 3)$. $h(\hat{\gamma}) < 0$ for $\hat{\gamma} = 1.15$ and $h(\hat{\gamma}) > 0$ for $\hat{\gamma} = 1.1$. Now we use simple bi-section method to get $\hat{\gamma} = 1.126458$ with $h(\hat{\gamma}) = -1.00156e^{-12}$. Using the above value of $\hat{\gamma}$ and (4.3.8), we get $\hat{\lambda} = 0.2631458$.

R-commands for computation are given in the appendix of this chapter.

(D) The Lognormal Distribution

Let $t_1, t_2, \cdots, t_n$ be a realisation of random sample of size n from log-normal distribution with parameter μ and σ^2.

The simplest way of obtaining the estimates of μ and σ^2 is by considering the normal distribution of $Y = \log_e T$, where T is lifetime random variable.

Thus, MLEs of μ and σ^2 are

$$\hat{\mu} = \frac{1}{n}\sum_{i=1}^{n} \log_e t_i$$

and

$$\hat{\sigma}^2 = \frac{1}{n}\left[\sum_{i=1}^{n}(\log t_i)^2 - \frac{\left(\sum_{i=1}^{n} \log t_i\right)^2}{n}\right],$$

by using well-known results for normal distribution.

The estimator of $\hat{\mu}$ is unbiased. However, the estimator of σ^2 is biased. An unbiased estimator of σ^2 is given by

$$\tilde{\sigma}^2 = \hat{\sigma}^2\left[\frac{n}{(n-1)}\right].$$

The maximum likelihood estimates of the mean and the variance of T, therefore, are

$$\exp(\hat{\mu} + \hat{\sigma}^2/2) \quad \text{and} \quad [e^{\hat{\sigma}^2} - 1][e^{2\hat{\mu}+\hat{\sigma}^2}]$$

respectively. $100(1-\alpha)\%$ confidence intervals for μ and σ^2 are;

$$\hat{\mu} - t_{\alpha/2}\frac{\hat{\sigma}}{\sqrt{n}} < \mu < \hat{\mu} + t_{\alpha/2}\frac{\hat{\sigma}}{\sqrt{n}}$$

where $t_{\alpha/2}$ is such that $P[T_{n-1} \geq t_{\alpha/2}] = \alpha/2$ where T_{n-1} follows student's t distribution with $(n-1)$ degrees of freedom. Similarly, the $100(1-\alpha)\%$ confidence interval for σ^2 is

$$\frac{n\hat{\sigma}^2}{\chi^2_{1-\alpha/2,n-1}} < \sigma^2 < \frac{n\hat{\sigma}^2}{\chi^2_{\alpha/2,(n-1)}}.$$

Our interest is in estimating the mean time to failure and confidence interval for average lifetime. Suppose mean time to failure is τ then MLE of $\tau(\hat{\tau})$ (using unbiased estimator of σ^2), is given by

$$\hat{\tau} = \exp\left[-\left(\hat{\mu} + \frac{\hat{\sigma}^2}{2} \times \frac{n}{n-1}\right)\right].$$

By Shapiro and Gross (1981), for large samples, $\hat{\tau}$ is approximately normal, with variance $\sigma^2_{\hat{\tau}}$, given as

$$\hat{\sigma}^2_{\hat{\tau}} = Var(\hat{\mu}) + \frac{1}{n} \cdot \frac{n^2}{(n-1)^2} Var(\hat{\sigma}^2).$$

The $100(1-\alpha)\%$ confidence interval for mean time to failure is

$$\exp[\hat{\tau} - Z_{1-\alpha/2}\hat{\sigma}_{\tau}] < \tau < \exp[\hat{\tau} + Z_{1-\alpha/2}\hat{\sigma}_{\tau}].$$

4.4 Parametric Analysis of Censored Data

In the last section we have discussed analysis of complete or uncensored data. In this section, we shall see how to apply similar techniques to censored data with appropriate modifications.

(i) Type I Censoring.

Let $X_1, X_2, ..., X_n$ be a random sample from the distribution $F_{\underline{\theta}}$ and t_0 be the fixed censoring time. What we observe are $t_1, t_2, ..., t_n$ where $t_i = x_i$ if $x_i \le t_0$, that is,

$$t_i = \begin{cases} x_i \text{ if } x_i \le t_0 \\ t_0 \text{ if } x_i > t_0 \end{cases}.$$

Let $R =$ the number of uncensored observation in the interval $(0, t_0]$.

Then R is a random variable with binomial distribution with parameters n and $p = F_{\underline{\theta}}(t_0)$. Therefore, its probability mass function (p.m.f.) is given by

$$f_R(n_u) = \binom{n}{n_u} p^{n_u}(1-p)^{n-n_u}, n_u = 0, 1, 2, ..., n.$$

Note that

$$P[R = 0] = [\overline{F}_{\underline{\theta}}(t_0)]^n.$$

The likelihood is

$$\begin{aligned} L(\underline{t}, \underline{\theta}) &= f_X[x_{(1)}, ..., x_{(n_u)}; \underline{\theta}/R = n_u] \times f_R(n_u), \quad \text{where} \\ &\quad \underline{t} = (x_{(1)}, \cdots, x_{(n_u)}, t_0, \cdots, t_0)' \\ &= n_u! \prod_{i=1}^{n_u} \left[\frac{f_\theta(x_{(i)})}{F_{\underline{\theta}}(t_0)} \right] \binom{n}{n_u} [F_{\underline{\theta}}(t_0)]^{n_u} [\overline{F}_{\underline{\theta}}(t_0)]^{n-n_u}, \\ &\quad 0 < x_{(1)} < x_{(2)} \cdots < x_{(n_u)} \le t_0 < \infty. \end{aligned}$$

$$L(\underline{t};\underline{\theta}) = \prod_{i=1}^{n_u} f_{\underline{\theta}}(x_{(i)}) \frac{n!}{(n-n_u)!} [\overline{F}_{\underline{\theta}}(t_0)]^{n-n_u}, 0 < x_{(1)} \leq ... \leq x_{(n_u)} < \infty.$$

Once we have the expression for the likelihood function, the likelihood based inference follows.

The Exponential Distribution

$$f_\lambda(x_{(i)}) = \lambda e^{-\lambda x_{(i)}}.$$
$$\overline{F}_\lambda(t_0) = e^{-\lambda t_0}.$$
$$L(\underline{t};\lambda) = \lambda^{n_u} e^{-\lambda \sum_1^{n_u} x(i)} \frac{n!}{(n-n_u)!} e^{-\lambda t_0 (n-n_u)}. \tag{4.4.1}$$
$$\frac{\delta}{\delta\lambda} \log L(\lambda) = \frac{n_u}{\lambda} - \left[\sum_{i=1}^{n_u} x_{(i)} + t_0(n-n_u)\right],$$
$$\hat{\lambda} = \frac{n_u}{\sum_{i=1}^{n_u} x_i + t_0(n-n_u)},$$

$$\hat{\lambda} = \frac{n_u}{T},$$

where T = the total time for which the n sample units are on test prior to the termination of the study.

$\hat{\mu}$ = estimate of the mean = $\frac{T}{n_u}, n_u > 0$. For $n_u = 0$, the estimate is not defined, but may be taken as nt_0.

From (4.4.1) it is clear that (T, n_u) is jointly sufficient statistic for λ. Theoretically, statistical procedures should be based on sufficient statistic. However, in this case it is difficult to do so as the dimension of the sufficient statistic is two whereas the dimension of the parameter space is one. Simple reasonably good procedures can be based on n_u alone, even though it takes into account only the number of failures and not the times of failures.

A point estimate of λ based on n_u alone is obtained by noting that R is binomial.

$$\tilde{p} = \frac{n_u}{n} = 1 - \exp[-t_0\tilde{\lambda}].$$

This gives an estimator

$$\tilde{\lambda} = \frac{-\log(1-\frac{n_u}{n})}{t_0}.$$

Bartholomew (1963) compares $\tilde{\lambda}$ and $\hat{\lambda}$ and derives the limiting efficiency;

$$\lim_{n\to\infty} \frac{Var\hat{\lambda}}{Var\tilde{\lambda}} = \frac{(1-p)(\log(1-p))^2}{p^2}.$$

As one might expect, the MLE is preferable to $\tilde{\lambda}$, however $\tilde{\lambda}$ is quite good for small p. Relative efficiency of $\tilde{\lambda}$ exceeds 96% for $p \leq 0.5$.

The principal advantage of R is its simple distributional properties. Clearly, confidence intervals or tests of hypotheses may be developed for λ by using known results for binomial distribution. For example, consider the test for $H_0 : \lambda = \lambda_0$ against $H_1 : \lambda < \lambda_0$. H_0 is rejected at $\alpha\%$ level of significance if n_u is too small, that is, if $B(n_u; n, p_0) \leq \alpha$, where $B(x; n, p)$ denotes cumulative binomial probability and $p_0 = 1 - \exp[-t_0\lambda_0]$.

Inferences based on $\hat{\mu}$: The distribution of $\hat{\mu}$ is complicated. Mendenhall and Lehman (1960) have studied the exact mean and variance of $\hat{\mu}$ for small samples. Bartholomew (1963) gives the exact (cumulative) distribution function of $\hat{\mu}$, given that $n_u > 0$, as a weighted sum of chi-square integrals. He considered different approximations for this distribution. In particular, given $n_u > 0, Z \xrightarrow{a} N(0,1)$ where

$$Z = \frac{u\sqrt{np}}{(1 - (2uq\log q)/p + qu^2)^{1/2}}, \tag{4.4.2}$$

$$u = \frac{(\hat{\mu} - \mu)}{\mu}, \quad p = 1 - e^{-t_0/\mu}, q = 1 - p.$$

The symbol $\xrightarrow{a}$ denotes "asymptotically distributed as".

This approximation is adequate for large p, $(p > 1/2)$. Recall that the procedures based on n_u are efficient for $p \leq 1/2$. Hence the following rule of thumb is suggested "Use the approximation given by (4.4.2) if $p_0 = 1 - e^{-t_0/\mu_0} = F(t_0)$ is larger than 0.5 otherwise base the inference on binomial distribution of R". This procedure is relatively convenient for testing of hypothesis problems but the confidence interval estimation is tedious.

It should be noted that the case $n_u = 0$ should not be completely ignored. If alternative hypothesis is $\mu > \mu_0$ then $n_u = 0$ should be included in the critical region. The rejection rule, in this case, is: Reject H_0 if $n_u = 0$ and for $n_u > 0$ reject H_0 if $Z_0 \geq z_{1-\alpha^*}$ where $\alpha^* = \alpha - \exp(\frac{-nt_0}{\mu_0})$ and Z_0 is observed (computed) value of the test statistic.

Illustration 4.4: (Bartholomew (1963)).

Suppose 20 items from an exponential distribution are put on life test and observed for 150 hrs. During the period 15 items fail with the following lifetimes, measured in hrs:

3, 19, 23, 26, 27, 37, 38, 41, 45, 58, 84, 90, 99, 109, 138.

Test the hypothesis $H_0 : \mu = 65$ against $H_1 : \mu > 65$ at 2.5% level of significance.

Solution:

$$p_0 = 1 - \exp(-\frac{t_0}{\mu_0}) = 1 - \exp(-\frac{150}{65}) = 0.9005.$$

As $p_0 > 0.5$, we use (4.4.2) to test the hypothesis.

$$\begin{aligned}
\hat{\mu} &= \frac{837 + 5(150)}{15} = 105.8. \\
\alpha &= 0.025 \\
\alpha^* &= 0.025, \quad z_{1-\alpha^*} = 1.96. \\
u_0 &= (\frac{\hat{\mu} - \mu_0}{\mu_0}) = 0.6277. \\
Z_0 &= \frac{u_0\sqrt{np_0}}{(1 - (\frac{2u_0 q_0 \log q_0}{p_0}) + q_0 u_0^2)^{1/2}} \\
&= \frac{2.6638}{(1 - (-0.3201) + 0.039)^{1/2}} = 2.2848.
\end{aligned}$$

$Z_0 > z_{1-\alpha^*}$. Therefore H_0 is rejected and we conclude that the average life is greater than 65 hours.

R-commands are given in the Appendix of this chapter.

Example 4.1: Derive the Fisher information based on one observation from an exponential distribution with type I censoring.

Solution: Define

$$\delta = \begin{cases} 1 \text{ if } x \leq t_0 \\ 0 \text{ if } x > t_0, \end{cases}$$

where t_0 is the fixed censoring time.

$$\begin{aligned}
L(x, \lambda) &= \lambda^{\delta} e^{-\delta\lambda x} . e^{-\lambda t_0(1-\delta)}. \\
\log(x; \lambda) &= \delta \log \lambda - \delta\lambda x - \lambda t_0(1 - \delta).
\end{aligned}$$

$$\frac{\delta}{\delta\lambda}\log(x;\lambda) = \frac{\delta}{\lambda} - \delta x - t_0(1-\delta).$$
$$E[-\frac{\delta^2}{\delta\lambda^2}\log l(\lambda)] = \frac{1}{\lambda^2}E(\delta) = \frac{1}{\lambda^2}[F(t_0)].$$
$$I(\lambda) = \frac{1}{\lambda^2}[1 - e^{-\lambda t_0}].$$

(ii) Type II Censoring.

In this case, the experimenter puts n units on test and decides to call off the experiment as soon as m failures have been observed. The data consist of the m smallest order statistics $X_{(1)}, X_{(2)}, ..., X_{(m)}$ (where m is fixed in advance) and the information that $(n-m)$ lifetimes are larger than $X(m)$. Let $f(x, \underline{\theta})$ be the probability density function (p.d.f.) of X. Then

$$L(\underline{t}, \underline{\theta}) = \binom{n}{m} m! \prod_{i=1}^{m} f(x_{(i)})[\overline{F}(x_{(m)})]^{n-m},$$

where $\underline{t} = (x_{(1)}, \cdots, x_{(m-1)}, x_{(m)}, \cdots, x_{(m)})'$

$$0 < x_{(1)} ... < x_{(m)} < \infty.$$

$$= \frac{n!}{(n-m)!} \prod_{i=1}^{m} f(x_{(i)})[\overline{F}(x_{(m)})]^{n-m},$$
$$0 < (x_{(1)} < ... < x_{(m)} < \infty.$$

(A) The Exponential Distribution: For exponential distribution with parameter λ (reciprocal of the mean),

$$L(x_{(1)}, x_{(2)}, ..., x_{(m)}, m \le n; \underline{\theta}) = \frac{n!}{(n-m)!}\lambda^m e^{-\lambda \sum_1^m x_{(i)}} e^{-\lambda(n-m)(x_{(m)})},$$
$$0 < x_{(1)} ... < x_{(m)} < \infty.$$

$$\log_e L(\lambda) \propto m \log\lambda - \lambda \sum_1^m x_{(i)} - \lambda(n-m)x_{(m)}.$$

$$\left[\frac{\delta}{\delta\lambda}\log L(\lambda)\right]_{\lambda=\hat{\lambda}} = 0 \Rightarrow$$

$$\hat{\lambda} = \frac{m}{\sum_{i=1}^{m} x_{(i)} + x_{(m)}(n-m)} = \frac{m}{T},$$

where T = the total time for which the n units were on test.

$$I(\lambda) = -E\left[\frac{\delta^2}{\delta\lambda^2}\log L(\lambda)\right] = \frac{m}{\lambda^2}.$$

Further we can write

$$\hat{\lambda} = \left[\frac{1}{m}\sum_{i=1}^{m} D_i\right]^{-1} = m\left[\sum_{i=1}^{m} D_i\right]^{-1}$$

where $D_i = (n - i + 1)(X_{(i)} - X_{(i-1)})$, the normalised sample spacings. So that

$$\hat{\mu} = \frac{1}{\hat{\lambda}} = \frac{\sum_{i=1}^{m} D_i}{m}$$

= Average of m i.i.d. exponential variables. This gives,

$$\frac{2m\lambda}{\hat{\lambda}} = \frac{2m\hat{\mu}}{\mu} \to \chi^2_{2m}.$$

$$E(\frac{1}{\hat{\lambda}}) = E(\hat{\mu}) = \mu.$$

Therefore $\hat{\mu}$ is an unbiased estimator of μ. Further $\hat{\mu}$ is the MVUE (Minimum Variance Unbiased Estimator) based on type II censoring scheme.
Remark: This is an unusual situation in which the censored results are very similar to the complete sample case. It is observed that all the results for complete case are valid after replacing n by m. It is clear that the statistical procedures based on these data are equivalent to the data obtained by putting m units on test and observing all m failures. A natural question is what are the advantages and disadvantages of the two sampling procedures? The principal advantage of type II censoring is that it may take less time for the first m failures to occur in a sample of size n, than for all m failures in a random sample of size m. A disadvantage is that additional $(n - m)$ items must be procured and put on test. Thus, the method of sampling which should be employed depends upon the relative cost of sampling and testing extra units.
Illustration 4.5

Sandhya and Dinesh are testing items from a population having exponential time to failure. Sandhya places six items on test and waits until they all fail. Dinesh, on the other hand, places ten items on test and discontinues the testing when the sixth failure occurs.

The expected time for Sandhya to complete her test is

$$\frac{1}{\lambda}[\frac{1}{6}+\frac{1}{5}+\frac{1}{4}+\frac{1}{3}+\frac{1}{2}+1]=\frac{49}{20\lambda}.$$

[Note that it is the time to failure for a parallel system composed of six identical components.]

The expected time for Dinesh to complete the test is

$$\frac{1}{\lambda}[\frac{1}{10}+\frac{1}{9}+\frac{1}{8}+\frac{1}{7}+\frac{1}{6}+\frac{1}{5}]=\frac{2131}{2520\lambda}.$$

(Note that it is the time to failure for a 5-out-of-10 system of identical exponential components.) The ratio of Dinesh's expected time to complete the test to Sandhya's expected time to complete the test is 0.345. So Dinesh can expect to finish his test about 65% sooner than Sandhya. The price that Dinesh pays for this time savings is in terms of costs related to the four additional items to be put on test. The cost of failed items is identical for Sandhya and Dinesh since six items fail in both cases. The four that survive Dinesh's test are as good as new by "no ageing" property of exponential distribution.

(B) The Gamma Distribution: Estimation of the parameters becomes considerably difficult for the gamma distribution in the presence of censoring.

For the gamma distribution with λ and γ as scale and shape parameters respectively, the likelihood of (λ, γ) given the type II censored sample is

$$\begin{aligned}
&L(x_{(1)}, x_{(2)}, ..., x_{(m)}, m \leq n; \lambda, \gamma) \\
&= \frac{n!}{(n-m)!}\prod_{i=1}^{m}\left\{\frac{\lambda^{\gamma}}{\Gamma(\gamma)}\exp(-\lambda x_{(i)})(x_{(i)})^{\gamma-1}\right\} \\
&\times\left[\int_{x_{(m)}}^{\infty}\frac{\lambda^{\gamma}}{\Gamma(\gamma)}x^{\gamma-1}e^{-\lambda x}dx\right]^{n-m}, \quad 0 < x_{(1)} < x_{(2)} ... < x_{(m)} < \infty. \quad (4.4.3)
\end{aligned}$$

Let

$$G' = \frac{[\Pi x_{(i)}]^{1/m}}{x_{(m)}}, A' = \left[\frac{\sum_{i=1}^{m} x_{(i)}}{m x_{(m)}}\right], \tau = \lambda x_{(m)}, f = \frac{m}{n}.$$

Note that, in terms of the above notation $x_{(m)}G'$ is the geometric mean and $x_{(m)}A'$ is the arithmetic mean of the complete observations.

$$L(\tau,\gamma) \propto \frac{\tau^{m\gamma} G'^{m(\gamma-1)})}{[\Gamma(\gamma)]^m} \times \exp[-(m\tau A')] \times [\int_1^\infty \frac{\tau^\gamma}{\Gamma(\gamma)} t^{\gamma-1} e^{-\tau t} dt]^{(n-m)}.$$

$$\log L(\tau,\gamma) \propto m\gamma \log\tau + m(\gamma-1)\log G' - m\tau A' + (n-m)\log[\int_1^\infty \frac{\tau^\gamma}{\Gamma(\gamma)} t^{\gamma-1} e^{-\tau t} dt] - m\log(\Gamma(\gamma))$$

$$\begin{aligned}\frac{\delta}{\delta\gamma}\log L(\tau,\gamma) &= m\log\tau + m\log G' - m\frac{\Gamma'(\gamma)}{\Gamma(\gamma)} + (n-m)\left\{\frac{\frac{d}{d\gamma}[\int_1^\infty \frac{\tau^\gamma}{\Gamma(\gamma)} t^{\gamma-1}e^{-\tau t}dt]}{\int_1^\infty \frac{\tau^\gamma}{\Gamma(\gamma)} t^{\gamma-1}e^{-\tau t}dt}\right\} \\ &= m\log\tau + m\log G' - m\frac{\Gamma'(\gamma)}{\Gamma(\gamma)} + (n-m)\frac{I'_\gamma(\gamma,\tau)}{I(\gamma,\tau)}\end{aligned}$$

and

$$\frac{\delta}{\delta\tau}\log l(\tau,\gamma) = \frac{m\gamma}{\tau} - mA' + (n-m)\frac{I'_\tau(\gamma,\tau)}{I(\gamma,\tau)}$$

(with obvious notation).

The MLEs of τ and $\gamma, \hat{\tau}$ and $\hat{\gamma}$, satisfy the equations

$$m\log\hat{\tau} + m\log G' - m\frac{\Gamma'(\hat{\gamma})}{\Gamma(\hat{\gamma})} + \frac{(n-m)I'_{\hat{\gamma}}(\hat{\gamma},\hat{\tau})}{I(\hat{\gamma},\hat{\tau})} = 0 \tag{4.4.4}$$

and

$$\frac{m\hat{\gamma}}{\hat{\tau}} - mA' + (n-m)\frac{I'_{\hat{\tau}}(\hat{\gamma},\hat{\tau})}{I(\hat{\gamma},\hat{\tau})} = 0. \tag{4.4.5}$$

Or

$$\log\hat{\tau} + \log G' - \frac{\Gamma'(\hat{\gamma})}{\Gamma(\hat{\gamma})} + \frac{(\frac{1}{f}-1)I'_{\hat{\gamma}}(\hat{\gamma},\hat{\tau})}{I(\hat{\gamma},\hat{\tau})} = 0 \tag{4.4.6}$$

and

$$\frac{\hat{\gamma}}{\hat{\tau}} - A' + (\frac{1}{f}-1)\frac{I'_{\tau'}(\hat{\gamma},\hat{\tau})}{I(\hat{\gamma},\hat{\tau})} = 0. \tag{4.4.7}$$

From (4.4.6) and (4.4.7) we see that the maximum likelihood equations are expressed only in terms of $\hat{\gamma}, \hat{\tau}, G', A'$ and f. Wilk et al. (1962) provide tables to aid in computing $\hat{\gamma}$ and $\hat{\lambda}$ for observed values G', A' and f. Bain and Engelhardt (1991) have also given tables to aid the computations of $\hat{\gamma}$ and $\hat{\tau}$.

(C) Lognormal Distribution: Type I / Type II censored data

Suppose failure times of $r(\leq n)$ units are available when n independent and identical units are on test. These are

$$t_{(1)} < t_{(2)} \cdots < t_{(r)}.$$

In case of type I censoring, these are the failure times of the r units which fail within the fixed interval $(0, t_0]$ of observation and in case of type II censoring, these are the failure times of fixed number r of failures. As in case of complete data we use the fact that $Y = \log_e T$ is $N(\mu, \sigma^2)$. We shall describe the method of Cohen (1959, 1961) for the estimation of μ and σ^2. It may be noted that the method is applicable when n is large.

Let

$$\overline{y} = \frac{1}{r}\sum_{i=1}^{r} \log_e t_{(i)}$$

and

$$s^2 = \frac{1}{r}\left[\sum_{i=1}^{r}(\log_e t_{(i)})^2 - \frac{\left(\sum_{i=1}^{r} \log_e t_{(i)}\right)^2}{r}\right].$$

The MLEs of μ and σ^2 are

$$\hat{\mu} = \overline{y} - \lambda(\overline{y} - \log_e t_{(r)}) \quad \text{and} \quad \hat{\sigma}^2 = s^2 + \lambda(\overline{y} - \log_e t_{(r)}).$$

The coefficient λ (Cohen, 1961) is a complicated function of α and β where

$$\alpha = \frac{s^2}{(\overline{y} - \log_e t_{(r)})^2} \tag{4.4.8}$$

and

$$\beta = \frac{(n-r)}{n} = \text{Proportion of censored units.} \tag{4.4.9}$$

Cohen (1961) provides tabulated values of λ as a function of α and β. Alternatively, λ can be calculated using the following approximation

$$\lambda = [1.136\alpha^3 - \log_e(1-\alpha)][1+0.437\beta - 0.25\alpha\beta^{1/3}] + 0.08[\alpha(1-\alpha)]. \quad (4.4.10)$$

This is good approximation for large values of n.

The asymptotic variances of $\hat{\mu}$ and $\hat{\sigma}$ can be estimated as

$$Var(\hat{\mu}) = m_1\hat{\sigma}^2/n,$$

$$Var(\hat{\sigma}) = m_2\hat{\sigma}^2/n,$$

$$cov(\hat{\mu}, \hat{\sigma}) = \frac{\hat{\sigma}^2}{n} m_3.$$

Cohen also provides tabulated values of m_1, m_2 and m_3 as a function of $\hat{c}$, where

$$\hat{c} = \frac{(\log_e t_{(r)} - \hat{\mu})}{\hat{\sigma}}.$$

Alternatively m_1 and m_2 can be approximately calculated as follows:

Let $y = -\hat{c}$.

For $y < 0$,

$$m_1 = 1 + 0.51e^{2.5y}$$
$$m_2 = 0.5 + 0.74e^{1.6y}.$$

For $y > 0$,

$$m_1 = 0.52 + e^{(1.8384y + 0.354y^2)} - 0.391y - 0.676y^2$$
$$m_2 = 0.24 + e^{(y + 0.384y^2)} + 0.2735y^2.$$

Illustration 4.6

Fifty units are subjected to fatigue test and the test is terminated when 35 units fail. Their lifetimes (in weeks) are given below:

22.3 26.8 30.3 31.9 32.1 33.3 33.7 33.9 34.7 36.1 36.4 36.5 36.6
37.1 37.6 38.2 38.5 38.7 38.7 38.9 38.9 39.1 41.1 41.1 41.4 42.4
43.6 43.8 44.0 45.3 45.8 50.4 51.3 51.4 51.5

Assume lognormal distribution and estimate the two parameters of the distribution. Also estimate

- The mean time to failure.

- The median time to failure and
- The standard deviation of time to failure.

Solution

The estimates of the two parameters of the distribution are:
$\mu = 3.809713$, $\sigma = 0.2808095$
The mean time to failure = 46.95266
The median time to failure = 45.13748
The standard deviation of time to failure = 13.44899.

In the above computations λ is computed using the approximation given by (4.4.8), (4.4.9) and (4.4.10).

R-commands for computations are provided in the appendix of this chapter.

(iii) *Random Censoring.*

Let $X_1, X_2, ..., X_n$ be the lifetimes of the n independent identical units on test. That is, $X_1, X_2, ..., X_n$ is a random sample from the distribution F. However, with each X_i, there is an associated random variable C_i, known as its censoring variable and therefore what we observe are $T_i = min(X_i, C_i), i = 1, 2, ..., n$ and

$$\delta_i = \begin{cases} 1 \text{ if } X_i \leq C_i \\ 0 \text{ if } X_i > C_i \end{cases}.$$

Let the censoring variable C_i have the p.d.f. g and distribution function G. It is assumed that T_i and C_i are independent random variables. Without this assumption, only few results are available. However, before applying the results which will be derived subsequently, one should carefully see whether the assumption of independence of T_i and C_i is justifiable. For example, in clinical trials when reason for withdrawal is related to the course of the disease, this assumption may not be satisfied.

Note that random censoring includes Type I censoring by setting $C_i \equiv t_0$, in which case the censoring distribution is degenerate at t_0.

The data consist of pairs $(t_i, \delta_i), i = 1, 2, ..., n$. The likelihood of the single pair viz. (t_i, δ_i) is

$$L(t_i, \delta_i : \underline{\theta}) = [f_{\underline{\theta}}(t_i)(1 - G(t_i)]^{\delta_i}[g(t_i)(1 - F_{\underline{\theta}}(t_i))]^{1-\delta_i}.$$

Therefore

$$L((t_1, \delta_1), ..., (t_n, \delta_n); \underline{\theta}) = \prod_1^n [L(t_i, \delta_i; \underline{\theta})]$$

$$= \prod_{t_i \in U} f_{\underline{\theta}}(t_i) \prod_{t_i \in C} \overline{F}_{\theta}(t_i) \times \prod_{t_i \in U} \overline{G}(t_i) \prod_{t_i \in C} g(t_i), \tag{4.4.11}$$

where U is the set of uncensored (complete) observations and C is the set of censored observations.

The last two terms in the likelihood, viz.

$$\prod_C g(t_i) \text{ and } \prod_U [1 - G(t_i)]$$

do not involve the unknown lifetime parameters because of the assumption that lifetime distribution and censoring distributions are independent. Hence these two terms are treated as constant while maximizing the likelihood.

(A) Exponential Distribution

$$L((t_1, \delta_1), ..., (t_n, \delta_n); \lambda) \propto \prod_{i \in U} \lambda e^{-\lambda t_i} \prod_{i \in C} e^{-\lambda t_i}.$$

$$\log L[(t_1, \delta_1), ..., (t_n, \delta_n); \lambda] \propto n_u \log \lambda - \lambda \sum_{i=1}^n t_i.$$

$$\frac{\delta}{\delta \lambda} \log L((t_1, \delta_1), ..., (t_n, \delta_n); \lambda) = \frac{n_u}{\lambda} - \sum_{i=1}^n t_i.$$

Therefore the likelihood equation:

$$\left[\frac{\delta}{\delta \lambda} \log L(\lambda) \right]_{\lambda = \hat{\lambda}} = 0 \Rightarrow \hat{\lambda} = \frac{n_u}{\sum_1^n t_i},$$

where n_u (the number of complete observation) as well as $(\sum_{i=1}^n t_i)$ are random variables.

Further, $i(\lambda) = \frac{n_u}{\lambda^2}$.

It may be noted that as type I censoring is a special case of random censoring with $c_i \equiv t_0$, for exponential distribution, as before, we get

$$\hat{\lambda} = \frac{n_u}{\sum_1^{n_u} t_{(i)} + (n - n_u)t_0}$$

which is also a ratio of two r.v.s.

Illustration 4.7 (Cox and Oakes (1984))

An experiment is conducted to determine the effect of a drug named 6-mercaptopurine (6-MP) on Leukemia remission times. A sample of $n = 21$ leukemia patients is treated with 6-MP and the times to remission are recorded. There are $r = 9$ individuals for whom the remission time is observed and the remaining 12 individuals are randomly right censored. Letting + denote the censored observation, the remission times (in weeks) are:

$4, 5, 6, 7^+, 8, 9^+, 10, 11^+, 12^+, 13, 16, 17^+, 19^+, 20^+, 22, 23, 25^+, 32^+, 33^+, 34^+, 35^+$.

Assuming exponential model,

$$\begin{aligned}\log L(\lambda) &= r \log \lambda - \lambda \sum_{i=1}^{n} t_i \\ &= 9 \log \lambda - 361\lambda.\end{aligned}$$

(Also see Figure 4.7.) This gives,

$$\hat{\lambda} = \frac{r}{\sum_{i=1}^{n} t_i} = \frac{9}{361} = 0.0249$$

or the MLE of the expected remission time is $= (.0249)^{-1} = 40.1606$.

The sample information at $\hat{\lambda}$ is

$$i(\hat{\lambda}) = \frac{(361)^2}{9} = 14480.1111.$$

Under the assumption that $2\lambda \sum_{i=1}^{n_u} t_i$ is approximately chi-square (this is exactly satisfied in case of complete and in the type II cases) an approximate $100(1 - \alpha)\%$ symmetric confidence interval for λ is

$$\left(\frac{\hat{\lambda}\chi^2_{2r,\alpha/2}}{2r}, \frac{\hat{\lambda}\chi^2_{2r,1-\alpha/2}}{2r} \right).$$

For $\alpha = 0.05$, it is given by $(0.014, 0.0437)$ for the data at hand.

Another interval estimator can be based on the likelihood ratio statistic, which is distributed asymptotically as a chi-square random variables with one degree of freedom. Thus with probability $(1-\alpha)$, the inequality

$$2[\log L(\hat{\lambda}) - \log L(\lambda)] \leq \chi^2_{1,1-\alpha}$$

is satisfied. For the particular example with $\alpha = 0.05$ this can be rearranged as

$$\log L(\lambda) \geq \log L(\hat{\lambda}) - \frac{3.84}{2}.$$

Or

$$\log L(\lambda) \geq -44.$$

Hence interval estimate is obtained by solving

$$9 \log \lambda - 361\lambda = -44.$$

The two roots of this equation obtained by plotting the graph (shown in Figure 4.6).

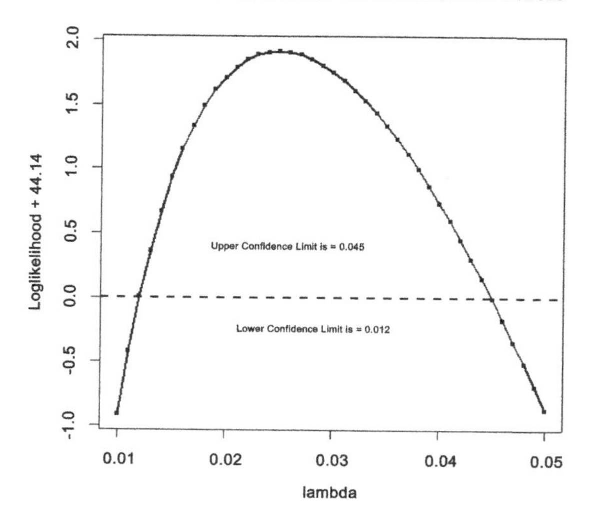

Figure 4.6

From the graph the required confidence interval is (0.0112, 0.045).

The final interval estimate of λ is based on the asymptotic normality of $\hat{\lambda}$. This is given by

$$\hat{\lambda} \pm (\hat{\lambda}/\sqrt{r}) \times z_{1-\alpha/2}.$$

For the problem at hand it is (0.008642, 0.041218).

Figure 4.7 shows the graph of log likelihood function.

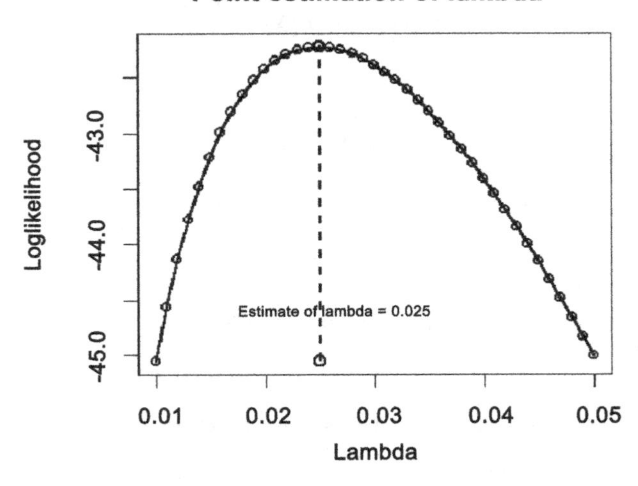

Figure 4.7

In view of the shape of the likelihood function a symmetric confidence interval would be inappropriate.

R-commands are given in the appendix of this chapter.

(B) Lehmann or Proportional hazards family.

Let

$$\overline{F}_\theta(x) = [\overline{F}_0(x)]^\theta, \quad \theta > 0, \quad \text{and} \quad f_\theta(x) = \theta(\overline{F}_0(x))^{\theta-1} f_0(x)$$

where $\overline{F}_0$ is a known survival function. $\overline{F}_\theta$ is a survival function depending on unknown parameter θ. The problem is to estimate θ under random censoring scheme. The data are

$$T_i = min(X_i, C_i) \text{ and } \delta_i = \begin{cases} 1 \text{ if } X_i \leq C_i \\ 0 \text{ if } X_i > C_i \end{cases}.$$

$$L((t_i, \delta_i), i = 1, 2, ..., n; \theta) \propto \prod_{i \in U} f_\theta(t_i) \prod_{i \in C} \overline{F}_\theta(t_i).$$

$$\begin{aligned}\log L(\theta) &\propto \sum_{i \in U} \log f_\theta(t_i) + \sum_{i \in C} \log \overline{F}_\theta(t_i) \\ &= \sum_{i \in U} \log \theta + (\theta - 1) \sum_{i \in U} \log \overline{F}_0(t_i) \\ &\quad + \sum_{i \in U} \log f_0(t_i) + \theta \sum_{i \in C} \log \overline{F}_0(t_i).\end{aligned}$$

$$\begin{aligned}\frac{\delta}{\delta\theta} \log L(\theta) &= \frac{n_u}{\theta} + \sum_{i \in U} \log \overline{F}_0(t_i) + \sum_{i \in C} \log \overline{F}_0(t_i) \\ &= \frac{n_u}{\theta} + \sum_{i=1}^{n} \log \overline{F}_0(t_i).\end{aligned}$$

Therefore,

$$\left[\frac{\delta}{\delta\theta} \log L(\theta)\right]_{\theta = \hat{\theta}} = 0 \Rightarrow \hat{\theta} = \frac{-n_u}{[\sum_{i=1}^{n} \log \overline{F}_0(t_i)]} \tag{4.4.12}$$

where $n_u = \#$ in U.

Observe that for complete sample ($n_u = n$) and

$$\hat{\theta} = \frac{-n}{\sum_{i=1}^{n} \log \overline{F}_0(t_i)}. \tag{4.4.13}$$

The following result (probability integral transformation) is useful for constructing confidence interval for θ:

Let $Y = F_0(X)$ then $Y \to U(0, 1)$ (that is distribution Y is uniform on the interval $[0, 1]$). Therefore

$$1 - Y = \overline{F}_0 \to U(0, 1).$$

Hence

$$H = -\log \overline{F}_0 \to \text{ Exponential } (1).$$

Therefore

$$P[H > x] = e^{-x}, x \geq 0.$$

Let $H_\theta = -\log \overline{F}_\theta$, then

$$P[H_\theta > x] = P[-\log \overline{F}_\theta > x] = P[-\log[\overline{F}_0] > x/\theta] = e^{-x/\theta}.$$

That is, $-\log_e \overline{F}_\theta \to \exp(1/\theta)$.

(C) Lognormal Distribution

Estimation of μ and σ^2 becomes more complicated. Interested readers are referred to Gajjar and Khatri (1969) and Cohen (1963, 1976).

(D) Weibull Distribution

As before, let $X_1, X_2, ..., X_n$ be the lifetimes and $C_1, C_2, ..., C_n$ be the associated censoring times. What we observe are $(t_i, \delta_i), i = 1, 2, ..., n$ where $t_i = min(X_i, C_i)$ and

$$\delta_i = \begin{cases} 1 \text{ if } X_i \leq C_i \\ 0 \text{ if } X_i > C_i \end{cases}.$$

For Weibull lifetime distribution,

$$\overline{F}(x) = \overline{e}^{\lambda x^\gamma};$$
$$f(x) = \lambda\gamma e^{-\lambda x^\gamma} x^{(\gamma-1)} \quad x \geq 0; \lambda,\ \gamma > 0.$$

$$L((t_1, \delta_1), (t_2, \delta_2), ..., (t_n, \delta_n); \lambda, \gamma) = L(\lambda, \gamma) \text{ say.}$$

Then,

$$\log L(\lambda, \gamma) \propto n_u \log \lambda + n_u \log \gamma + (\gamma - 1) \sum_{i \in U} \log t_i - \lambda \sum_{i \in U} t_i^\gamma - \lambda \sum_{i \in C} t_i^\gamma$$

$$= n_u \log \lambda + n_u \log \gamma + (\gamma - 1) \sum_{i \in U} \log t_i - \lambda \sum_{1}^{n} t_i^\gamma.$$

$$\frac{\delta}{\delta\lambda} \log L = \frac{n_u}{\lambda} - \sum_{i=1}^{n} t_i^\gamma$$

$$\frac{\delta}{\delta\gamma} \log L = \frac{n_u}{\gamma} + \sum_{i \in U} \log t_i - \lambda \sum_{i=1}^{n} t_i^\gamma \log t_i.$$

So, the MLE is of λ and γ, say $\hat{\lambda}$ and $\hat{\gamma}$ satisfy

$$\hat{\lambda} = \frac{n_u}{\sum\limits_{i=1}^{n} t_i^{\hat{\gamma}}} \tag{4.4.14}$$

and

$$\frac{n_u}{\hat{\gamma}} + \sum_{i \in U} \log t_i - \hat{\lambda} \sum_{i=1}^{n} t_i^{\hat{\gamma}} \log t_i = 0. \qquad (4.4.15)$$

These equations have no closed form solutions for $\hat{\lambda}$ and $\hat{\gamma}$. However, the first equation gives $\hat{\lambda}$ in terms of $\hat{\gamma}$. Using this expression for $\hat{\lambda}$, the second equation yields a single but complicated expression with $\hat{\gamma}$ as the only unknown in the implicit equation:

$$h(\hat{\gamma}) = \frac{n_u}{\hat{\gamma}} + \sum_{i \in U} \log t_i - \frac{n_u \sum_{i=1}^{n} t_i^{\hat{\gamma}} \log t_i}{\sum_{i=1}^{n} t_i^{\hat{\gamma}}} = 0.$$

This equation is solved iteratively by using numerical methods such as Newton-Raphson method. The initial or starting solution for γ is obtained by a graphical procedure similar to the one described for complete case with slight modification. $\log\log[\overline{F}_n(x_{(i)})]^{-1}$ is plotted against $\log x_{(i)}$ for uncensored observations ($t_i = x_i$) and $\log \lambda_0$ and γ_0 are estimated by the y intercept and the slope of the line thus obtained. Alternatively one can use Kaplan-Meier estimator (discussed in Chapter 5) of survival function.

The sample information matrix $i(\lambda, \gamma)$ is

$$i(\lambda, \gamma) = \begin{bmatrix} \frac{n_u}{\lambda^2} & \sum_{1}^{n} t_i^{\gamma} (\log t_i) \\ & \frac{n_u}{\gamma^2} + \lambda \sum_{i=1}^{n} t_i^{\gamma} (\log t_i)^2 \end{bmatrix}.$$

The expected value of $i(\lambda, \gamma)$ is *not* mathematically tractable. If one desires to estimate (λ, γ) simultaneously, Fisher's method of scoreing cannot be used. However, one can use the Newton-Raphson method of scoring by taking the initial solution as given by the graphical method.

Example 4.2: We derive the scores test for exponentiality against IFR alternative if the lifetime distribution is Weibull distribution.

Let $T \rightarrow W(\lambda, \gamma)$. The testing problem is:

$$H_0 : \gamma = 1, H_1 : \gamma > 1.$$

The score is $[\frac{\delta}{\delta\gamma}\log L(\lambda,\gamma)]_{\gamma=1,\lambda=\hat{\lambda}} = U_{\gamma_0}$ say.

$$U_{\gamma_0} = n_u + \sum_{i\in U}\log t_i - \hat{\lambda}\sum_{i=1}^{n} t_i \log t_i$$

$$= n_u + \sum_{i\in U}\log t_i - (\frac{n_u}{\sum_1^n t_i}).\sum_{i=1}^{n} t_i \log t_i.$$

The sample information matrix at $(\gamma_0, \hat{\lambda})$ has elements

$$I_{\gamma\gamma} = n_u + \left(\frac{n_u}{\sum t_i}\right)\sum_{i=1}^{n} t_i(\log t_i)^2.$$

$$I_{\gamma\lambda} = \sum_1^n t_i \log t_i.$$

$$I_{\lambda\lambda} = \frac{(\sum_1^n t_i)^2}{n_u}.$$

The inverse matrix V has leading element

$$V_{\gamma\gamma} = (I_{\gamma\gamma} - \frac{I_{\gamma\lambda}^2}{I_{\lambda\lambda}})^{-1}.$$

The signed statistic $U_{\gamma_0}(V_{\gamma\gamma})^{1/2}$ is approximately a standard normal variable. Thus the test will reject for large values of the above standardized statistic with reference to the appropriate critical value from the standard normal distribution.

Illustration 4.8

Remission times for leukemia patients for 6-MP group are: 6, 6, 6, 7, 10, 13, 16, 22, 23, 6+, 9+, 10+, 11+, 17+, 19+, 20+, 25+, 32+, 32+, 34+, 35+

Assume Weibull model and estimate the parameters.

Solution: We use graphical method to get crude estimates of the parameters. Figure 4.8 is the plot of $\log(t_{(i)})$ vs. $\log[-\log(S(t_{(i)}))]$.

Estimate of parameter gamma from graphical method is 1.24259. We shall use iterative method to improve this crude solution.

Final estimate of gamma is 1.353735 and lambda is 0.008528222.

R-commands are given in the appendix of this chapter.

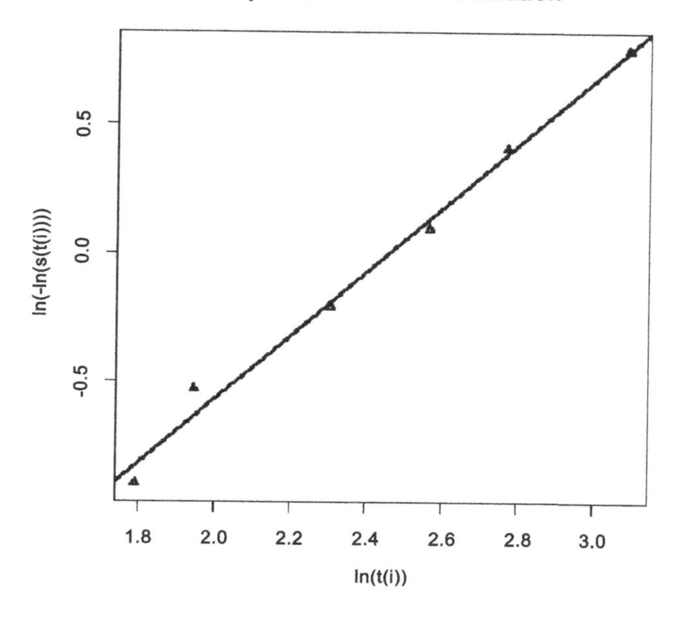

Figure 4.8

Exercises

Exercise 4.1: Suppose that 21 patients with acute leukemia have the following remission times in months:

$$1, 1, 2, 2, 3, 4, 4, 5, 5, 6, 8, 8, 9, 10, 10, 12, 14, 16, 20, 24.$$

(A) Assume that remission times follow an exponential distribution. Obtain the point estimate of the scale parameter and also interval estimate of scale parameter for 95% confidence coefficient.
(B) Carry out crude graphical test for exponentiality based on empirical and estimated survival functions.
Exercise 4.2: The density function of two-parameter exponential distribution is:

$$f(t) = \begin{cases} \lambda \exp(-\lambda(t - G)) & t \geq G \geq 0, \quad \lambda \geq 0 \\ 0 & t < G \end{cases}$$

where W is called the guarantee time, or the minimum survival time before which no deaths occur. Obtain survival function.

Consider the survival times in months of 11 patients following initial pulmonary metastasis from ostenogenie sarcoma considered by Burdette and Gehan (1970). The data were: 11, 13, 13, 13, 13, 13, 14, 14, 15, 17. Assume a two-parameter exponential distribution and estimate hazard rate after estimating the guarantee time by smallest observation in the data. Estimate the probability of surviving beyond 18 months.

Exercise 4.3: A mechanical engineer conducts a fatigue test by subjecting ten identical steel rods to a stress level higher than endurance level of the rod material. The number of cycles to failure are recorded as:

$$20000, 35000, 47000, 58000, 68000, 77000, 85000, 92000, 97000, 102000.$$

Assume that the cycles to failure follow gamma distribution. Estimate the parameters of the distribution and mean life.

Exercise 4.4: A manufacturer of an automotive speed sensor subjects 10 sensors to a reliability test that simulates the environmental conditions (temperature and speed) at which the sensors are normally operated. A sensor is classified failed when its output falls outside 5% tolerance. The number of miles accumulated before the failure of the sensores are:

$$110000, 130000, 150000, 155000, 159000, 163000, 166000, 168000, 169000, 170000.$$

Assume that the miles to failure follow Rayleigh distribution. Determine the parameter of the distribution, the mean life of sensor and the variance of its lifetime.

Exercise 4.5: Simulate 1000 observations from Weibull distribution with shape parameter 2 and scale parameter 5. Fit Weibull model to this data set.

Exercise 4.6: A production engineer performs a burn-in test on 8 video display terminals. The following failure times (in hrs) are recorded: 20, 28, 35, 39, 42, 44, 46, 47. Assume lognormal distribution of failure times and estimate mean failure time and its standard deviation.

Exercise 4.7: A manufacturer of end mill cutters introduces a new ceramic cutter material. In order to estimate the expected life of a cutter, the manufacturer places 10 units under continuous test and monitors the tool wear. A failure of the cutter occurs when ware-out exceeds a predetermined value. Because of the time constraint, the manufacturer decides to run the test for 50000 minutes. The times to failure are shown below:

$$3000, 7000, 12000, 18,000, 20000, 30000.$$

The distribution of failure times is exponential. Determine the mean life of a cutter made from this material.

What is the reliability at 60000 minutes?

Exercise 4.8: Suppose X denotes the life in years of a certain component and X has exponential distribution. 20 units were put on test and all the failures which occurred in the first 2 years were recorded. Suppose the following 10 values were recorded.

$$0.497, 0.638, 0.703, 0.839, 0.841, 0.950, 1.054, 1.103, 1.125, 1.495.$$

(a) Compute MLE of μ, the mean of the distribution.
(b) Test $H_0 : \mu = 2.9$ against $H_1 : \mu < 2.9$ at $\alpha = 0.01$.
(c) Find lower 99% confidence limit for μ.

Exercise 4.9: As an alternative to an automobile air-bag crash testing, a test engineer develops a sensor test system that uses a mechanical, vibration shaker to replay measured actual crashes. The sensors are subjected to the same conditions measured during a crash test. Ten sensors are placed under test for 50 hours. The following failure times are observed:

$$10, 20, 30, 35, 39, 42, 44, 50+, 50+, 50+.$$

Assume Rayleigh distribution and determine the parameter of the distribution and mean life of sensor.

Exercise 4.10: Fifty identical units are put on life test and the test is terminated after 10 of them have failed. The lifetimes of the failed units are:

$$65, 110, 380, 420, 505, 580, 650, 840, 910 \text{ and } 950.$$

If the lifetime of the component is exponentially distributed then estimate mean life of the component and its failure rate. Also calculate 90% confidence interval for mean lifetime.

Exercise 4.11: The following data shows the failure times of tires in thousands of miles:

$$3, 4, 6, 9, 9, 11, 12, 14, 16, 18, 30, 35, 38, 8+, 13+, 22+, 28+, 36+, 45+, 46+.$$

The + sign indicates censored observation. Assume that the distribution of failure times is Weibull. Use graphical method to estimate the parameter gamma of the distribution. Use iterative method to improve the estimate of gamma and estimate the parameter lambda.

Exercise 4.12: Twenty-one units are subjected to fatigue test. The times to failure in hours are:

$$8, 8, 8, 9, 13, 15, 18, 25, 26, 8+, 13+, 19+, 22, 29+, 33+, 36+, 40+, 45+, 47+, 49+.$$

Assuming exponential distribution of failure times, estimate the fatigue rate and the mean time to failure. Also obtain confidence interval for mean time to failure under the assumption that $2\lambda\Sigma t_i$ is exponentially distributed.

Appendix

R-Commands for Illustration 4.1: Exponential distribution (complete data)

We enter the data using c(combine) function.

```
> t <-c(840,157,145,44,33,121,150,280, 434, 736, 584, 887,
  263,1901,695, 294, 562, 721, 76, 710, 46, 402, 194, 759,
  319, 460, 40,1336,335, 1334, 454, 36, 667, 40, 556, 99,
  304, 375, 567, 139, 780, 203,436, 30, 384, 129, 9, 209,
  599, 83, 832, 328, 246, 1617, 638,937, 735, 38, 365, 92,
  82, 220);
> # The vector, t, of waiting times;

> n <-length(t); # The number,n, of components of vector t;
> n;     # Print value of n;
[1] 62
> estlam<-n/sum(t) # Estimate of lambda;
> estlam   # Print value of estimate of lambda;
[1] 0.002288921
> LCL<-estlam*2*qgamma(.025,shape=n)/(2*n);
> # Lower confidence limit (LCL);
>  LCL;    # Print LCL;
[1] 0.001754903
> UCL<-estlam*2*qgamma(.925,shape=n)/(2*n);
> # Upper confidence limit (UCL);
> UCL;     # Print value of UCL;
[1] 0.002719646
```

Graphical Methods for checking Exponentiality:

1. We plot empirical and estimated survival curves on the same graph paper. If the two curves are close then the model is appropriate.

```
> d<-c(rep(1,62));       # Vector of the frequency;
> cd<-cumsum(d);         # Vector of cumulative frequency;
> emps<-(n-cd)/n;        # Empirical survival function;
> t<-sort(t);            # Ordered vector of observations;
> s<-exp(-(estlam*t));   # Estimate of the survival function;
```

Plotting commands:

```
> plot(t,emps,"o", pch = 1, lwd=2,xlab="Waiting time",
  ylab="Survival time",main="survival functions",cex=.7);
> # Plot empirical survival function;
```

The first two arguments of the command plot () are x, y coordinates. The third argument, l, means line graph, fourth argument states the width of the line, arguments 5, 6, 7 are titles, the last argument controls the font size.

```
> points(t,s,"o",pch=2);
> # Add the plot of empirical survival function;
```

The fourth argument of the function points() is point type.

```
> legend (locator(1), legend = c ("empirical",
  "estimated exponential"),pch=1:2,cex=0.7);
> # Adds the legend after clicking at the appropriate place;
> # in the graph;
```

2. We plot -log(S(t)), where S(t) is empirical survival function versus t. If data are from exponential distribution, the graph will show the linear trend.

```
> x<-t;
> y<- ( -log(emps));
> y<-y[-62];
> # Assign to object y all the elements of original vector;
> # y but last element;
> x<-x[-62];
```

```
> # Assign to object x all the elements of original vector;
> # x but last element;
```

Plotting commands:

```
> plot(x,y,"o",pch=1,lwd=2,xlab="Time",ylab="-log(S(t))",
  main="Test for Exponentiality",cex=.7);
```

Commands to plot line of best fit and the equation of line of best fit along with the test for significance of regression coefficient, coefficient of correlation etc.

```
> abline(lm(y~x), lwd=2) # Add the line of best fit;
> legend (locator (1), legend = " -log (S(t))", pch = 1, cex = 0.7);
> summary(lm(y ~ x));

Call:
lm(formula = y ~ x)

Residuals:
     Min        1Q    Median        3Q       Max
-0.25617 -0.07655  0.00731  0.06571  0.45486
Coefficients:
              Estimate Std. Error t value Pr(>|t|)
(Intercept) -7.120e-02  2.658e-02  -2.679  0.00955 **
x            2.516e-03  4.901e-05  51.333  < 2e-16 ***
---
Signif. codes:  0 *** 0.001 ** 0.01 * 0.05 . 0.1   1

Residual standard error: 0.1346 on 59 degrees of freedom
Multiple R-squared:  0.9781,     Adjusted R-squared:  0.9777
F-statistic:  2635 on 1 and 59 DF,  p-value: < 2.2e-16

> # Output showing summary of the fit of the ;
> # linear model to the data;
```

Illustration 4.2: Gamma Distribution (complete data)

We enter the data by using scan( ) function.

```
> x<-scan()
```

```
1: 370 1055 1270 1502 1763 706 1085 1290 1505 1768
11: 716 1102 1293 1513 1781 746 1102 1300 1522 1782
21: 785 1108 1310 1522 1792 797 1115 1313 1530 1820
31: 844 1120 1315 1540 1868 855 1134 1330 1560 1881
41: 858 1140 1355 1567 1890 886 1199 1390 1578 1893
51: 886 1200 1416 1594 1895 930 1200 1419 1602 1910
61: 960 1203 1420 1604 1923 988 1222 1420 1608 1940
71: 990 1235 1450 1630 1945 1000 1238 1452 1642 2023
81: 1010 1252 1475 1674 2100 1016 1258 1478 1730 2130
91: 1018 1262 1481 1750 2215 1020 1269 1485 1750 2268
101: 2440
102:
Read 101 items
> a<-mean(x);   # Computation of arithmetic mean (a.m.);
> lg<-sum(log(x,10))/length(x);
> #  Computation of log geometric mean (g.m.);
> g <-10 ^(lg);     # Computation of g.m.;
> a;g;     # Print a.m. and g.m ;
[1] 1400.911
[1] 1342.259
>  r<-g/a; r;    # Ratio of g.m. to a.m.;
[1] 0.9581333
> 1/(1-r);
[1] 23.88532
> y1<-20; y2<-30; gam1<-9.91125; gam2<-14.91305; y<-23.88532;
> # Input the values from tables of Wilk et. al. (1962);
> gam<-gam2-((y2-y)/(y2-y1)*(gam2-gam1)); #Linear interpolation;
> gam;     # Estimate of gamma;
[1] 11.85461
> lam<-gam/a; lam;     # Estimate of Lambda;
[1] 0.008462072
```

Plotting commands

```
> x1<-sort(unique(x)) # Ordered array of distinct failure epochs;
> y1<-pgamma(x1,shape=gam,scale=(1/lam));
> # Estimated distribution function;
> plot(x1,y1,"l",lty=2,lwd=2,
  main="Empirical and Estimated Cumulative Distribution",cex=.6);
```

```
>  xt<-table(x) # Prepare frequency table;
> n <-length (x);
> y2<- cumsum(xt)/n # Empirical distribution function;
>  lines(x1,y2,"s",lwd=2);
> legend(locator(1),legend=c("Empirical","Estimated Gamma"),
  lty=c(1,2));
```

Weibull Distribution (Complete Data)

```
> t<-c(.35, .59, .96, .99, 1.69, 1.97, 2.07, 2.58, 2.71, 2.90,
  3.67, 3.99, 5.35, 13.77);    # Vector of failure epochs;

> n<-length(t);    # Number of components in t;
> d<-c(rep(1,n)); # Vector of number of failures;
> cd<-cumsum(d);   # Cumulative number of failures;
> s<-(n-cd)/n;     # Empirical survival function;
> s<-s[1:n-1];     # Vector of all but last component of s;
> t<-t[1:n-1];     # Vector of all but last component of t;
> x<-log(t);
> y<-log(-log(s));
```

Plotting Commands

```
> plot(x,y,xlab="log(t(i))",ylab="ln(-ln(s(t(i)))",
  main="Graphical Method of Estimation",cex=.7,lwd=2,"p",pch=2);
> abline(lm(y ~ x),"o",pch=1,lwd=2);
> text(locator(1),"Line of best fit and the data points");
> summary(lm(y ~ x)) #Give the summary of the fit of linear model;
Call:
lm(formula = y ~ x)
Residuals:
     Min       1Q   Median       3Q      Max
-0.22494 -0.09671  0.01793  0.08730  0.19642
Coefficients:
            Estimate Std. Error t value Pr(>|t|)
(Intercept) -1.27262    0.04703  -27.06 2.05e-11 ***
x            1.29732    0.04877   26.60 2.46e-11 ***
---
Signif. codes:  0 *** 0.001 ** 0.01 * 0.05 . 0.1   1
```

```
Residual standard error: 0.1341 on 11 degrees of freedom
Multiple R-squared:  0.9847,    Adjusted R-squared:  0.9833
F-statistic: 707.6 on 1 and 11 DF,  p-value: 2.462e-11
```

From graphical method the crude estimate of gamma is 1.29732. In order to get refined estimate we use first, approximate trial and error method. For this we compute the values of the function h(.) for values of gamma in the range (1,3).

```
> g<-seq(1,3,.05);
> # Values of gamma in the range (1,3) with increment of 0.05;
> length(g); # Number of components of vector g ;
[1] 41
> h<-1:41;
> t<-c(t,13.77);
> s1<-sum(log(t));
> i<-1;
> while(i<42)
 {
 h[i]<-((n / g[i]) +s1-(n/sum(t ^g[i])) *(sum(t ^g[i]*log(t))))
 i<-i+1
 }
> # Computation of vector of function h(g);
> h[1:20] # Print first 20 values of vector h;
 [1]   2.9007405   1.7046017   0.5741565  -0.4978862  -1.5174785
 [6]  -2.4895054  -3.4179962  -4.3062917  -5.1571768  -5.9729871
[11]  -6.7556961  -7.5069851  -8.2283019  -8.9209076  -9.5859164
[16] -10.2243269 -10.8370488 -11.4249237 -11.9887422 -12.5292574
> h[3] # Access 3rd value of vector h;
[1] 0.5741565
> h[4] # Access 4th value of vector h;
[1] -0.4978862
> g[3] # Print 3rd value of vector g;
[1] 1.1
> g[4] # Print 4th value of vector g;
[1] 1.15
```

For gamma = 1.1, h(gamma)>0 and for gamma=1.15, h(gamma) is negative. Hence for simple bi-section method we start with x1=1.1 and x2=1.15.

```
> x1<-1.1;x2<-1.15;x<-(x1+x2)/2;
> i<-1;
> while(i<20)
 {
  h<-((n/x+s1-(n/sum(t ^ x))*(sum(t ^x * log(t)))))
  if (h>0)
  x1<-x else x2<-x
  x=(x1+x2)/2
  i=i+1
  }
> # Loop to carry out 20 iterations of bisection method
> x1;x2;x;h; # Print X1, X2, estimate of gamma and value of;
> # function at this estimate;
[1] 1.126458
[1] 1.126458
[1] 1.126458
[1] -1.031472e-06
```

We have considered only 20 iterations and we have the estimate correct up to 12 decimal spaces.

```
> g <- x; # Assign to g the gamma estimate, x;
> l<-n/sum(t ^ g); # Obtain lambda;
> l; # Print estimate of lambda;
[1] 0.2631458
```

Alternate method of fitting of parametric Weibull distribution to survival data:
We use survreg function. It may be noted that:
There are multiple ways to parametrise a Weibull distribution. The survreg function imbeds it in a general location-scale family, which is a different parametrisation than the R-Weibull function and often leads to confusion. Hence it should be remembered that:
1. survreg's scale = 1/(R-Weibull shape)
2. survreg's intercept = log(R-Weibull scale).
Alternate Solution:

```
> y<-c(0.35, 0.59, 0.96, 0.99, 1.69, 1.97,  2.07,
     2.58, 2.71, 2.90, 3.67, 3.99, 5.35, 13.77);
> library(survival);
```

```
> survreg(Surv(y)~1, dist="weibull");
Call:
survreg(formula = Surv(y) ~ 1, dist = "weibull")
Coefficients:
(Intercept)
   1.185172
Scale= 0.8877382
Loglik(model)= -29.7   Loglik(intercept only)= -29.7
n= 14
```

The form of the distribution used by R is:
F(x) = 1 - exp(-(x/b)^a) on x > 0 where b is scale parameter and a is shape parameter.
weibull shape = 1/survrreg(scale) = 1/0.8877382 = 1.126458
weibull scale = exp(-survreg intercept / survregscale) = exp(-1.185172 / 0.8877382)= 0.2631459

Illustration 4.4: Exponential Distribution (Type I censored data)

```
> t<-c(3,19,23,26,27,37,38,41,45,58,84,90,99,109,138);
> # Vector of failure times
> n<-20;      # Sample size;
> r<- length (t);     r;     # Number of failures;
[1] 15
> mu0<-65;       # Specified value of mean;
> alpha<-0.025;       # Level of significance;
> t0<-150;

> p0<-1-exp(-t0/mu0); p0;      # Specified proportion;
[1] 0.9005094
 > estmu<-(sum(t)+(n-r)*t0)/r; # Estimated value of population mean;
> estmu;
[1] 105.8

> alpha1<-alpha-exp(-n*t0/mu0);     # Level of significance when;
> # zero is included in the critical region;
> alpha1;     # Print adjusted level of significance;
[1] 0.025
> qnorm((1-alpha1));
[1] 1.959964
```

```
> # Cutoff point for normal distribution for;
> # adjusted level of significance;
> u0<-(estmu - mu0)/mu0;
> u0;
[1] 0.6276923
> nz0<-u0*(n*p0)^.5;
> nz0;
[1] 2.663826
> dz1<-2*u0*(1-p0)*log(1-p0)/p0;
> dz1;
[1] -0.3200725
> dz2<-(1-p0)*u0^2;
> dz2;
[1] 0.03919905
> z0<-nz0/(1-dz1+dz2)^.5;  # Observed value of test statistic;
> z0;
[1] 2.284824
```

Illustration 4.6: Lognormal distribution (Type II censored data)

```
> x<-scan()
1: 22.3 26.8 30.3 31.9 32.1 33.3 33.7 33.9 34.7 36.1 36.4 36.5
13: 36.6 37.1 37.6 38.2 38.5 38.7 38.7 38.9 38.9 39.1 41.1 41.1
25: 41.4 42.4 43.6 43.8 44.0 45.3 45.8 50.4 51.3 51.4 51.5
36:
Read 35 items
> y<-log(x); # Vector(y) of log of failure times;
> r<-length(y); # Number of components(r) of y;
> r; # Print r;
[1] 35
> ymean<-mean(y); # Mean of log of failure times;
> ymean; # Print mean log time;
[1] 3.647393
> ss<-1/r*(sum((y-ymean)^2));
> ss;
[1] 0.03110124
> alpha<-ss/(ymean-y[r])^2; # Computation of coefficient alpha;
> alpha;
[1] 0.3593563
> n<-50;
```

```
> beta<-(n-r)/n; # Computation of coefficient of beta;
> beta;
[1] 0.3
> lambda<-(1.136*(alpha)^3-log(1-alpha)) *(1+0.437*beta-0.25*alpha*
  (beta)^(1/3))+0.08*alpha*(1-alpha);
> # Computation of coefficient of lambda;
> lambda;
[1] 0.5517543
> mu<-ymean-lambda*(ymean-y[r]); # Estimate of parameter mu;
> mu;     # Print mu;
[1] 3.809713
> sqsigma<-ss+lambda*(ymean-y[r])^2;
> sqsigma;
[1] 0.07885395
> sigma<-sqrt(sqsigma); sigma; # Estimate of parameter sigma;
[1] 0.2808095
> exp(mu)-> median; # Estimate of median;
> median;       # Print median;
[1] 45.13748
> m<-(x[25]+x[26])/2;
> m ;      # Sample median;
[1] 41.9
> ave<-exp(mu+1/2*sqsigma);
> ave;
[1] 46.95266
> sddev<-sqrt((exp(sqsigma)-1)*(exp(2*mu+sqsigma)));
> # Estimate of standard deviation;
> sddev;       # Print estimate of standard deviation;
[1] 13.44899
```

Illustration 4.7: Exponential distribution(Randomly(Right) Censored data)

```
> tu<-c(4,5,6,8,10,13,16,22,23); # Vector of failure times;
> r<-length(tu); r;              # Number of failures;
[1] 9
> tc<-c(7,9,11,12,17,19,20,25,32,33,34,35);
> # Vector of censored times;
> length(tc);                   # Number of censored observations;
[1] 12
```

```
> n<- length (c(tu, tc));
> estlam<-r/(sum(tu)+sum(tc)); # Computation of estimate of lambda;
> estlam;     # Print estimate of lambda;
[1] 0.02493075
> estmu<-1/estlam;     # Computation of estimate of population mean;
> estmu;     # Print estimate of population mean;
[1] 40.11111
```

Estimation of confidence interval under the assumption that $2\lambda \sum_{i=1}^{n_u} t_i$ is approximately chi-square.

```
> LCI<-estlam*2*qgamma(.025,shape=r)/(2*r);
> # Lower confidence limit;
> LCI;    # Print lower confidence limit;
[1] 0.01139993
> UCI<-estlam*2*qgamma(.975,shape=r)/(2*r);
> # Upper confidence imit;
> UCI;    # Print upper confidence limit;
[1] 0.04366534
```

Note: Lower confidence limit can also be obtained by the command

```
> LCL <-(estlam/(2*r)) * qchisq (df = 2 * r, 0.025);
```

and upper confidence limit by the command

```
> UCL <-(estlam/(2*r)) * qchisq (df = 2 * r, 0.975);
```

Estimation of confidence interval using likelihood ratio statistic

```
> x<-seq(.01,.05,.001);
> y<-9*log(x)-361*x+44.14;
```

Plotting commands for the graph

```
> plot(x,y,"o",xlab="lambda",ylab="Loglikelihood + 44.14",
  main="CI based on Likelihood ratio",lwd=2,cex=.7);
> options(warn=-1);
> abline(0,0,"l",lty=2,lwd=2);
> x[3];x[36];
[1] 0.012
[1] 0.045
```

```
> text(locator(1),"Lower Confidence Limit is=0.012",cex=.7);
> text(locator(1),"Upper Confidence Limit is=0.045",cex=.7);
```

Estimation of confidence interval based on asymptotic normality of $\hat{\lambda}$;

```
> LCI <-estlam-(estlam/r^0.5)*qnorm(0.975);
> LCI;
[1] 0.008642959
> UCI <-estlam+(estlam/r^0.5)*qnorm(0.975);
> UCI;
[1] 0.04121854
```

Plotting commands for graphical estimation of lambda

```
> x<-seq(.01,.05,.001);
> y<-9*log(x)-361*x; # Computation of log likelihood;
> plot(x,y,"o",xlab="lambda", ylab= "Loglikelihood ",
  main="Point Estimation of lambda",lwd=2,cex=.7);
> m<-which(y==max(y));
> est=x[m]; # Estimate of lambda;
> est;      # Print estimate of lambda;
[1] 0.025
> x1<-c(x[m],x[m]);
> y1<-c(min(y),max(y));
> lines(x1,y1,"o",lty=2,lwd=2);
> text(locator(1),"Estimate of lambda is=0.025",cex=.7);
```

Illustration 4.8: Weibull Distribution Randomly (Right) Censored Data.

Graphical Method:

```
> t<-c(6, 7, 10, 13, 16, 22, 23); # Vector of failure times;
> d<-c(3, 1, 1, 1, 1, 1, 1); # Vector of number of failures;
> n<-sum(d);
> cd<-cumsum(d); # Vector of cumulative number of failures;
> s<-(n-cd)/n; # Empirical survival function for uncensored data;
> x<-t[1:length(t)-1]; # Vector of all but last failure epoch;
> s<-s[1:length(s)-1]; # Vector of all but last survival time;
> y<-log(-log(s));
> x<-log(t[1:length(t)-1]);
> plot(x,y,"p",pch=2,lwd=2,xlab="ln(t(i))",ylab="ln(-ln(s(t(i))))",
```

```
  main="Graphical method of estimation",cex=.7);
> abline(lm(y ~ x));
> summary(lm(y~x)); # Output showing summary of the linear fit;

Call:
lm(formula = y ~ x)
Residuals:
       1        2        3        4        5        6
-0.07363  0.10616 -0.01523 -0.03762  0.01850  0.00181
Coefficients:
            Estimate Std. Error t value Pr(>|t|)
(Intercept) -3.05552    0.15129  -20.20 3.55e-05 ***
x            1.24259    0.06166   20.15 3.58e-05 ***
---
Signif. codes:  0 *** 0.001 ** 0.01 * 0.05 . 0.1   1
Residual standard error: 0.06834 on 4 degrees of freedom
Multiple R-squared:  0.9902,    Adjusted R-squared:  0.9878
F-statistic: 406.1 on 1 and 4 DF,  p-value: 3.579e-05
```

Estimate of parameter gamma from graphical method is 1.24259.
We shall use iterative method to improve this crude solution.

```
> tu<-c(rep(6,3),7,10,13,16,22,23);#Vector of complete observations;

> tc<-c(6,9,10,11,17,19,20,25,32,32,34,35);
> # Vector of censored observations;
> nu<-length(tu);nc<-length(tc);
> # Number of complete and censored observations;
> g<-seq(1,2,.05);
> length(g);
[1] 21
> h<-1:21;
> s1<-sum(log(tu));
> t <-c(tu, tc);
> i<-1;
> while(i<22)
  {
  h[i]<-(nu/g[i])+s1-(nu/sum(t^g[i]))*(sum(t^g[i]*log(t)))
  i<-i+1
```

```
  }
> # Compute function h(.);
> h[1:20]; # Print first 20 values of function h(.);
 [1]  3.18062138  2.62564870  2.11267650  1.63663263  1.19328311
 [6]  0.77906331  0.39094825  0.02635181 -0.31695226 -0.64089293
[11] -0.94715181 -1.23720332 -1.51234726 -1.77373536 -2.02239307
[16] -2.25923756 -2.48509269 -2.70070149 -2.90673658 -3.10380908

>  h[8];h[9];g[8];g[9]; # Print value of the function for the;
> # 8-th and 9-th value of the parameter;
[1] 0.02635181
[1] -0.3169523
[1] 1.35
[1] 1.4

> x1<-1.35; x2<-1.4;x=(x1+x2)/2;

> i<-1;
> while(i<20)
  {
  h<-((nu/x+s1-(nu/sum(t^x))*(sum(t^x*log(t)))));
  if (h>0)
  x1<-x else x2<-x;
  x=(x1+x2)/2;
  i=i+1;
  }
> x1;x2;x;h; #Print X1, X2, solution x & value of function at X;
[1] 1.353734
[1] 1.353735
[1] 1.353735
[1] 2.152312e-07

> t<-c(tu,tc); t; # Vector of all observations;
 [1]  6 6 6 7 10 13 16 22 23 6 9 10 11 17 19 20 25 32 32 34 35

> estlam<-nu/(sum(t^1.353735));  estlam; # Estimate of lambda
[1] 0.008528222
```

Chapter 5

Non-parametric Estimation of the Survival Function

5.1 Introduction

In the previous chapters, we have considered parametric models for life distribution and methods of estimation of the unknown parameters involved in these models. These methods are discussed for complete as well as censored data. Such a parametric modelling is based on prior knowledge of the failure characteristics of the individual (or unit). However, in many practical situations such prior knowledge may not be available. For example; (i) a production process is set up for the manufacture of a new item and (ii) a study is undertaken of a new virus or a new disease. Furthermore, the inferences based on parametric models may not be robust, in the sense that if any of the assumptions implicit in the choice of the model and/or in the methods used for the inferences, are not satisfied then the conclusions drawn therefrom are far from valid. For these reasons, in such situations, the non-parametric approach is advocated.

5.2 Uncensored (complete) Data

Let $X_1, X_2, ..., X_n$ be a random sample of size n from a distribution F. A non-parametric estimator of $F(x)$ is $F_n(x)$ where $F_n(x)$ is sample (or empirical) distribution function defined as

$$F_n(x) = \frac{1}{n} \text{ [The number of observations } \leq x]$$

The Glivenko-Cantelli theorem tells us that

$$\sup_x [|F_n(x) - F(x)|] \rightarrow 0 \text{ as } n \rightarrow \infty \quad \text{with probability 1.}$$

Hence, $F_n(x)$ as a function is a consistent estimator of $F(x)$. Then obviously, a non-parametric consistent estimator of survival function $[\overline{F}(x)]$ is $\overline{F}_n(x)$ where $\overline{F}_n(x)$ is defined by

$$\overline{F}_n(x) = \frac{\#(\text{observations} \;\; > x)}{n}.$$

It is seen that for a fixed x, $n\overline{F}_n(x)$ follows the binomial distribution with parameters n and $\overline{F}(x)$. Hence the standard asymptotic theory leads to the asymptotic normality of $\sqrt{n}(\overline{F}_n(x) - \overline{F}(x))$. Unbiasedness of this estimator is obvious. Further, using these results, confidence interval for $\overline{F}(x)$, for fixed x, can be constructed. Confidence bands for the entire $\overline{F}(x), 0 < x < \infty$ may be constructed using the distribution of $\sup_x |\overline{F}_n(x) - \overline{F}(x)|$.
Illustration 5.1 (Leemis, (1995)): A complete data set of $n = 23$ ball bearing failure times to test endurance of deep groove ball bearings has been extensively studied. The ordered set of failure times measured in 10^6 revolutions is

17.88, 28.92, 33.0, 41.52, 42.12, 45.60, 48.48, 51.84, 51.96, 54.12, 55.56, 67.80, 68.64, 68.64, 68.88, 84.12, 93.12, 98.64, 105.12, 105.84, 127.92, 128.04, 173.40.

Observe that the data set contain two observations tied at 68.64.

An empirical survival function has a downward step of $\frac{1}{n}$ at each observed lifetime. It is also the survival function corresponding to a discrete random variable for which n mass values are equally likely. Ties are not difficult to adjust for since the formula remains the same and the function will take a downward step of d/n if there are d tied observations at a particular time point. In Figure 5.1 the non-parametric estimator of $\overline{F}(t)$ is shown where the downward steps in $\overline{F}_n(t)$ are connected by vertical lines.

A point estimate $[S_n(t)]$ of the survivor function at t $(\overline{F}(t))$, for $t = 50$ (in the units of 10^6) is given by, $S_n(50) = \frac{16}{23} = 0.696$. Approximate 95% confidence interval for $S(t)$ is $S_n(50) \pm 1.96\sqrt{\frac{S_n(50)(1-S_n(50))}{23}}$ which reduces to

$$0.508 < S(50) < 0.884.$$

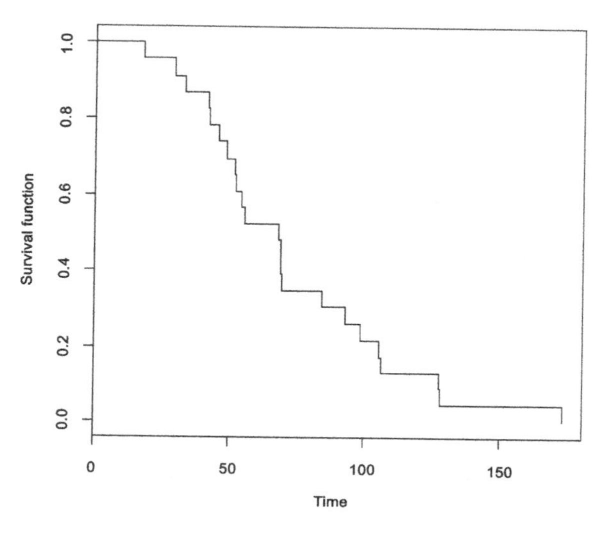

Figure 5.1

R-commands to plot empirical survival function are given in the appendix of this chapter.
Confidence bands: The confidence bands for the survival function can be obtained by using Kolmogorov-Smirnov statistic:

$$D_n = \sup_x |F_n(x) - F(x)|.$$

Therefore

$$\sup_x |\overline{F}_n(x) - \overline{F}(x)| = D_n.$$

From the tables of D_n statistic, find $D_{n,(1-\alpha)}$ such that

$$P[D_n \leq D_{n,(1-\alpha)}] = 1 - \alpha.$$

That is, $P[\sup_x |\overline{F}_n(x) - \overline{F}(x)| \leq D_{n,(1-\alpha)}] = 1 - \alpha$. This gives

$$P[\overline{F}_n(x) - D_{n,(1-\alpha)} \leq \overline{F}(x) \leq \overline{F}_n(x) + D_{n,(1-\alpha)} \quad \forall \ x] = 1 - \alpha.$$

As $0 \leq \overline{F}(x) \leq 1$,

$$L_n(x) = \max[\overline{F}_n(x) - D_{n,(1-\alpha)}, 0]$$

and

$$U_n(x) = min[\overline{F}_n(x) + D_{n,(1-\alpha)}, 1]$$

give the required confidence band.

The asymptotic probability distribution of D_n is an infinite sum, which may, in practice, be approximated by its first term $2e^{-2d^2}$.

That is, $\lim_{n\to\infty} P[D_n > d/\sqrt{n}] = 2e^{-2d^2}$. Setting this equal to α and solving we get

$$\lim_{n\to\infty} P[\hat{S}(t) - \frac{d_{(1-\alpha)}}{\sqrt{n}} < S(t) < \hat{S}(t) + \frac{d_{(1-\alpha)}}{\sqrt{n}}]$$
$$= 1 - \alpha.$$

Then $\hat{S}_n(t) \pm \frac{d_{(1-\alpha)}}{\sqrt{n}}$ define an asymptotic $100(1-\alpha)\%$ non-parametric confidence band for $S(t)$ for $t \leq T_{(n)}$. In Dixon and Massey (1983), the tables for $d_{(1-\alpha)}/\sqrt{n}$ are available. Nair (1984) has shown that the asymptotic critical value of $d_{(1-\alpha)}/\sqrt{n}$ is valid for n as small as 25.

5.3 Censored Data

The general case of randomly right censored data will be considered here.

(A) *The Actuarial Method*: Life tables have historically been used by actuaries for estimating the survival distribution of humans, but apply well to reliability and biostatistical situations for which grouped data (rather than raw data) which display the combined survival experience of a cohort of individuals who fall into natural groupings by age or calender time interval are available.

Notation: Suppose the time interval $(0, \tau]$ is under consideration. Let this be partitioned into a fixed sequence of intervals $I_1, ..., I_k$. These intervals are almost always, but not necessarily of equal lengths and for human populations the length of each interval is usually one year.

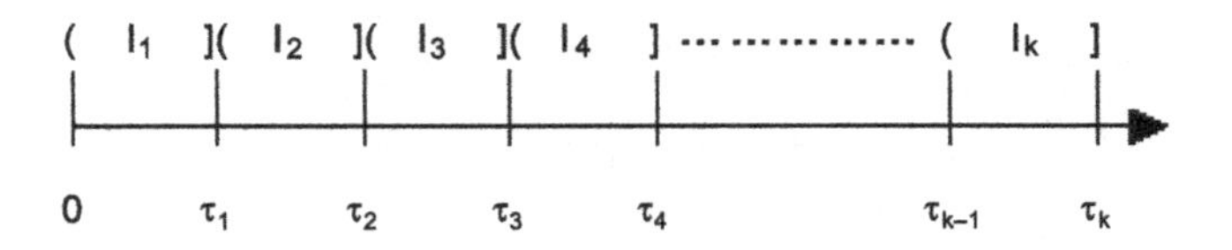

For a life table, let

$n_i = \#$ alive at the beginning of I_i,
$d_i = \#$ died during I_i,
$\ell_i = \#$ lost to follow-up during I_i,
$w_i = \#$ withdrawn during I_i,
$P_i = P$ [surviving through I_i / alive at the beginning of I_i]
$Q_i = 1 - P_i$

Table 5.1 (Cutler and Ederer (1958), Millar (1981)] is an example of a life table. In this table, Cutler and Ederer reviewed annual cohorts of Connecticut residents with localised kidney cancer diagnosed in the years 1946 to 1951. The study terminated on Dec. 31, 1951. Within each annual cohort, the patients were subdivided by years after diagnosis, commonly called as "time-on-study". For each such time interval, two groups of patients were defined: those who died during the interval and those who were lost to follow-up. (The latter group might also include deaths from other causes). During the last time-on-study interval of each cohort, a third category was defined; those known to be alive at the end date of study. The term used to describe these patients was "withdrawn alive". In our terminology, patients lost to follow-up or withdrawn alive are said to have censored survival times.

Table 5.1: Life table

Year after diagnosis	Alive at the beginning of interval	Died during the interval	Lost to followup during the Interval	Withdrawn alive during the interval
0 - 1	126	47	4	15
1 - 2	60	5	6	11
2 - 3	38	2	-	15
3 - 4	21	2	2	7
4 - 5	10	-	-	6

We break up the survival probability $S(\tau_k)$ into a product of conditional probabilities:

$$\begin{aligned} S(\tau_k) &= P[T > \tau_k] \\ &= P[T > \tau_1]P[T > \tau_2/T > \tau_1] \cdots P[T > \tau_k|T > \tau_{k-1}] \\ &= P_1.P_2...P_k, \end{aligned}$$

where

$$P_i = P[T > \tau_i | T > \tau_{i-1}].$$

The actuarial method estimates P_i seperately and then multiplies the estimates to get the estimate of $S(\tau_k)$.

For an estimate of P_i we would have used $(1 - \frac{d_i}{n_i})$ if the data were complete, i.e. there were no losses and withdrawals in I_i. We assume that, on average, those individuals who are lost to follow-up or withdrawn during I_i were at risk for half of the interval. Therefore, the effective sample size is defined as

$$n_i' = n_i - \frac{1}{2}(\ell_i + w_i)$$

and

$$\hat{Q}_i = q_i = \frac{d_i}{n_i'}, \quad \hat{P}_i = p_i = 1 - \hat{Q}_i.$$

Then the actuarial estimate of $S(\tau_k)$ is

$$\hat{S}(\tau_k) = \pi_{i=1}^{k} \hat{P}_i.$$

Illustration 5.2 (Ebeling (1997)): The following table shows the annual failures and removals (censored) of a fleet of 200 single engine aircrafts. Removals resulted from aircrafts eliminated from the inventory for various reasons other than engine failure.

The following table shows the life table for engine failure data.

Table 5.2

Year	Working at the beginning of interval n_i	Failed during the interval d_i	Censored during the interval $(\ell_i + w_i)$
1981	200	5	0
1982	195	10	1
1983	184	12	5
1984	167	8	2
1985	157	10	0
1986	147	15	6

Table 5.2 Cont'd

Year	Working at the beginning of interval n_i	Failed during the interval d_i	Censored during the interval $(\ell_i + w_i)$
1987	126	9	3
1988	114	8	1
1989	105	4	0
1990	101	3	1

Note: Year 1980 is taken as base year.

Table 5.3

Acturial estimate of reliability function

τ_i	n_i	$\hat{p}_i$	$\hat{S}(\tau_i)$: Reliability at τ_i
1	200	0.975	0.975
2	194.5	0.949	0.925
3	181.5	0.934	0.864
4	166	0.952	0.823
5	157	0.936	0.770
6	144	0.896	0.690
7	124.5	0.928	0.640
8	113.5	0.930	0.595
9	105	0.962	0.573
10	100.5	0.970	0.556

$$\hat{S}(\tau_k) = 0.556$$

For R-commands see the appendix of this chapter.
Variance of $\hat{S}(\tau_k)$: To estimate the variance of $\hat{S}(\tau_k)$ we use the delta method.
Delta Method: This method is used to obtain the approximate variance of a function $g(Y)$ in terms of the variance of Y.
Univariate Case: Suppose Y is a r.v. with mean μ and variance of σ^2 and we want the expression for the mean and the variance of $g(Y)$. Assuming

differentiability of g, expand $g(Y)$ about μ.

$$g(Y) = g(\mu) + (Y - \mu)g'(\mu) + \frac{(Y-\mu)^2}{2!}g''(\mu) + ...$$

This gives $E[g(Y)] = g(\mu)$ by ignoring the second and higher order terms. Further we have approximately,

$$E[g(Y) - g(\mu)]^2 = E(Y-\mu)^2[g'(\mu)]^2 = \sigma^2[g'(\mu)]^2.$$

If $Y \overset{a}{\sim} N[\mu, \sigma^2]$, then

$$g(Y) \overset{a}{\sim} N(g(\mu), \sigma^2(g'(\mu))^2).$$

Now we shall apply this method to obtain an approximate expression for $Var(\hat{S}(\tau_k))$. Consider

$$\log \hat{S}(\tau_k) = \sum_{i=1}^{k} \log \hat{P}_i,$$

we can see that $n_i'\hat{P}_i$ is approximately distributed as $B(n_i', P_i)$ [this is suggested by Epstein and Sobel (1953)].

$$Var(\log \hat{P}_i) \doteq Var(\hat{P}_i)\left[\left(\frac{d}{d\hat{P}_i}(\log \hat{P}_i)\right)^2\right]_{\hat{P}_i = P_i} \doteq \frac{P_iQ_i}{n_i'} \times \frac{1}{P_i^2}.$$

Ignoring the covariances between $\log \hat{P}_1, ..., \log \hat{P}_k$ we can write

$$\begin{aligned} est.Var[\log \hat{S}(\tau_k)] &= \sum_{i=1}^{k} \frac{\hat{Q}_i}{n_i'\hat{P}_i} \\ &= \sum_{i=1}^{k} \frac{d_i}{n_i'(n_i' - d_i)}. \end{aligned}$$

Now, using the delta method again we get

$$est.Var(\hat{S}(\tau_k)) = [\hat{S}(\tau_k)]^2 \sum_{i=1}^{k} \frac{d_i}{n_i'(n_i' - d_i)}.$$

The above formula for the estimate of variance of $\hat{S}(\tau_k)$ is known as *Greenwood's Formula.*

For the life table data quoted above Table 5.4 shows estimates of standard error of $\hat{S}(\tau_i)$ by Greenwood's formula:

Table 5.4
Estimate of standard error of $\hat{S}(\tau_i)$ using Greendwood's formula

τ_i	Standard Error $(\hat{S}(\tau_i))$
1	0.011
2	0.019
3	0.024
4	0.027
5	0.030
6	0.033
7	0.035
8	0.036
9	0.036
10	0.036

For R-commands see the appendix of this chapter.

The life-table analysis fails to account adequately for the advantage of the single-clinic retrospective study where the dates of entry and dates of deaths are known exactly. In such studies, the elapsed time from entry to the specified closing date for each still surviving patient is known precisely, so also for a patient lost to observation or dead from other causes. That is, censoring times are also known precisely. Grouping results in the loss of this information. The Kaplan-Meier method of estimating survival probabilities makes use of this information.

(B) *Product-Limit (Kaplan-Meier) Estimator*

Let $X_1, X_2, ..., X_n$ be the lifetimes of the n individuals (units). With each X_i there is associated a random variable C_i, known as its censoring variable. What we observe is $T_i = min(X_i, C_i)$ and

$$\delta_i = \begin{cases} 1 \text{ if } X_i \leq C_i \\ 0 \text{ if } X_i > C_i \end{cases}.$$

For the moment assume that there are no ties. Let $T_{(1)} < T_{(2)} \cdots < T_{(n)}$ be the order statistics corresponding to $T_1, T_2, ..., T_n$ and with a little abuse of notation, define $\delta_{(i)}$ to be the value associated with $T_{(i)}$. That is, $\delta_{(i)} = \delta_j$ if $T_{(i)} = T_j$. Let $R(t)$ denote the risk set at time t, which is the set of subjects still alive at time t^- (just prior to t) and $n_i = \#R(T_{(i)}) = \#$ alive at $T_{(i)}^-$, and $d_i = \#$ died at $T_{(i)}$.

It may be noted that for the data with no ties $d_i = 1$ or 0 depending on $\delta_{(i)}$ is 1 or 0.

The time interval of interest, in this case, is $(0, \tau]$ where $\tau = T_{(n)}$. We consider the subdivision of this interval into n subintervals I_i with end points $T_{(i)}$.

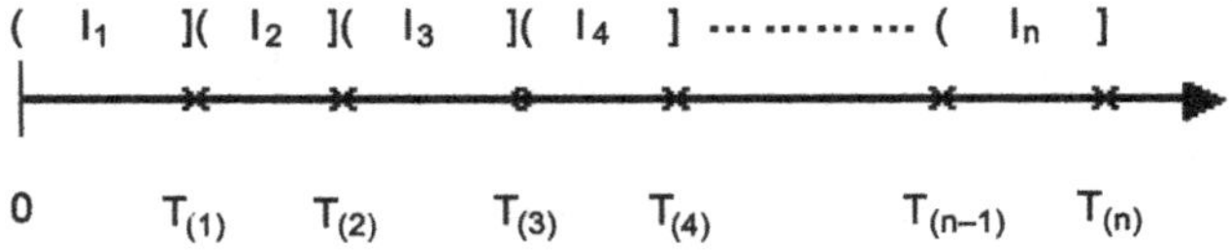

(I_1](I_2](I_3](I_4] .... (I_n]

0 $T_{(1)}$ $T_{(2)}$ $T_{(3)}$ $T_{(4)}$ $T_{(n-1)}$ $T_{(n)}$

On the time axis, $\times$ denotes $\delta_{(i)} = 1$, i.e. an uncensored observation and 0 denotes $\delta_{(i)} = 0$, i.e. censored observation.

Let $P_i = P$ [surviving through I_i / alive at the beginning of I_i]
$= P[T > T_{(i)} | T > T_{(i-1)}]$, and
$Q_i = 1 - P_i$. Its estimator is then $\hat{Q}_i = q_i = \frac{d_i}{n_i}$ and hence that of P_i is

$$\hat{P}_i = p_i = 1 - q_i = \begin{cases} 1 - \frac{1}{n_i} \text{ if } \delta_{(i)} = 1 \text{ (uncensored)} \\ \quad 1 \quad \text{if } \delta_{(i)} = 0 \text{ (censored)} \end{cases}$$

The PL (product limit) estimator of the survival function when no ties are present is then

$$\begin{aligned} \hat{S}(t) &= \prod_{U; T_{(i)} \le t} (1 - \frac{1}{n_i}) \\ &= \prod_{T_{(i)} \le t} (1 - \frac{1}{n_i})^{\delta_{(i)}} \\ &= \prod_{T_{(i)} \le t} (1 - \frac{1}{n - i + 1})^{\delta_{(i)}} \\ &= \prod_{T_{(i)} \le t} (\frac{n - i}{n - i + 1})^{\delta_{(i)}} \end{aligned}$$

[Kaplan and Meier (1958)].
Notes: (i) For tied uncensored observations, the factor for d deaths in the product - limit estimator is $(1 - \frac{d}{m})$ where m is the number at risk at the time point at which multiple deaths occur. Thus, it is not difficult to adjust

for ties. The form of the formula remains the same. The only difference is that $d > 1$.

(ii) If censored and uncensored observations are tied, consider the uncensored observation to occur before the censored observation.

(iii) If the last (ordered) observation $T_{(n)}$ is censored, then for $\hat{S}(t)$ as defined above,

$$\lim_{t\to\infty} \hat{S}(t) > 0.$$

Sometimes it is preferable to redefine $\hat{S}(t) = 0$ for $t \geq T_{(n)}$ or to think of it as being undefined for $t \geq T_{(n)}$ if $\delta_{(n)} = 0$.

Using notes (i) to (iii), by letting $T'_{(1)} < T'_{(2)} \cdots < T'_{(r)}$ denote the distinct survival times,

$$\delta'_{(j)} = \begin{cases} 1 \text{ if the observations at } T'_{(j)} \text{ are uncensored} \\ 0 \text{ if censored,} \end{cases}$$

$$n'_j = \#R(T'_{(j)}) \quad \text{and} \quad d_j = \# \text{ died at time } T'_{(j)}$$

the PL estimator allowing for ties is then

$$\hat{S}(t) = \prod_{U;T'_{(j)}\leq t} (1 - \frac{d_j}{n'_j}) = \prod_{T'_{(j)}\leq t} (1 - \frac{d_j}{n'_j})^{\delta'_{(j)}}.$$

Even after a span of more than 45 years of the development of the Kaplan-Meier estimator, it is still commonly used in clinical studies. Even though this estimator is non-parametric in nature it may be noted that the assumption of independence of lifetime distribution and censoring time distribution is essential.

Illustration 5.3: The following failure and censor times (in operating hrs) were recorded on 12 turbine vanes: 142, 149, 320, 345$^+$, 560, 805, 1130$^+$, 1720, 2480$^+$, 4210$^+$, 5280, 6890 (+ indicates censored observation). Censoring was a result of failure mode other than wearout. Plot PL estimate of survival function.

Table 5.5
Estimation of survival function for turbine vanes

j	$\tau_{(j)}$	d_j	n_j	$\hat{S}(\tau_{(j)})$
1	142	1	12	0.9167
2	149	1	11	0.8334
3	320	1	10	0.7500
4	345^+	0	–	**
5	560	1	8	0.6563
6	805	1	7	0.5625
7	1130^+	0	–	**
8	1720	1	5	0.4500
9	2480^+	0	–	**
10	4210^+	0	–	**
11	5230	1	2	0.2250
12	6890	1	1	0.0000

** These are censored observations, hence we have not computed survival probability for them. However, it may be noted that in the relevant intervals the survival probability remains constant.

Table 5.6
R-output of estimation of survival function for vanes

time	n.risk	n.event	survival
142	12	1	0.917
149	11	1	0.833
320	10	1	0.750
560	8	1	0.656
805	7	1	0.563
1720	5	1	0.450
5230	2	1	0.225
6890	1	1	0.000

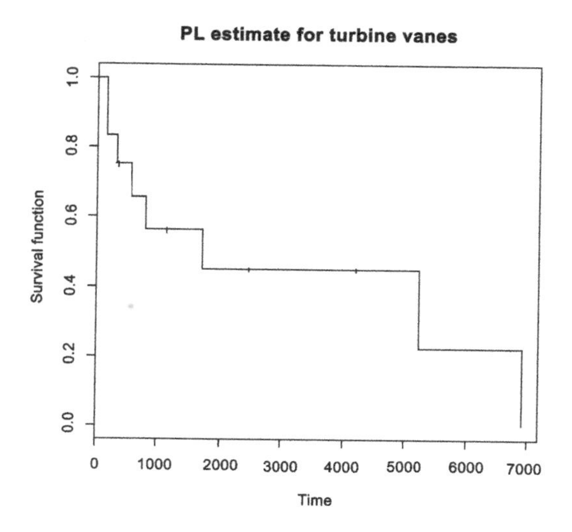

Figure 5.2

For R-commands to plot PL estimator for non-maintained group see the appendix of this chapter.

Illustration 5.4: A clinical trial to evaluate the efficacy of maintenance chemotherapy for acute myelogenous Leukemia (AML) was conducted by Embury et al. at Stanford University. After reaching a state of remission through treatment by chemotherapy, the patients who entered the study were randomised into two groups. The first group received maintenance chemotherapy; the second or control group did not. The objective of the trial was to see if maintenance therapy prolonged the time until relapse, that is, increased the length of remission.

For a preliminary analysis during the course of the trial the data were as follows:

Length of complete remission (in weeks).

Maintained group

9, 13, 13^+, 18, 23, 28^+, 31, 34, 45^+, 48, 161^+

Non-maintained group

5, 5, 8, 8, 12, 16^+, 23, 27, 30, 33, 43, 45.

It is often useful to plot survival functions of the two groups on the same plot so that they can be directly compared. Figure 5.3 shows the survival curves for the two groups.

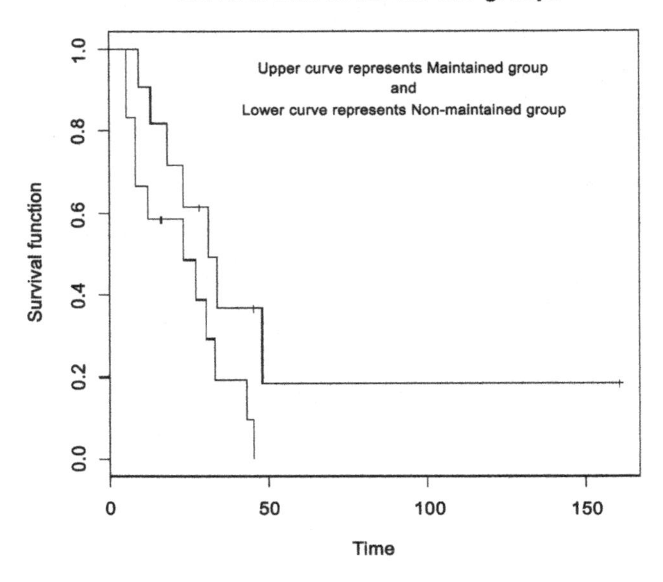

Figure 5.3

Variance of $\hat{S}(t)$: Using the same arguments as for the variance of the actuarial estimate, we get

$$\hat{Var}(\hat{S}(t)) = \hat{S}(t)^2 \sum_{T_{(i)} \leq t} \frac{\hat{Q}_i}{n_i \hat{P}_i}$$

$$= \hat{S}(t)^2 \sum_{T_{(i)} \leq t} \frac{\delta_{(i)}}{(n-i)(n-i+1)},$$

and with ties present the expression becomes

$$\hat{Var}(\hat{S}(t)) = \hat{S}^2(t) \sum_{T'_{(j)} \leq t} \frac{\delta'_{(j)} d_j}{n'_j(n'_j - d_j)}.$$

This formula is also called the Greenwood formula. The justification for these formulae is not as obvious as in case of life tables because the number of terms in the product is random. Using Greenwood's formula, the approximate standard error of $\hat{S}(24)$ in the maintained group is 0.1526.

For R-commands see the appendix.

In case of censored data, the confidence bands for $S(t)$ at values of t greater than the earliest censored time will be wider than that given by the Kolmogorov theory. Moreover, the band will continue to widen at later times as more and more censored times are encountered. We shall briefly outline and illustrate the method of Hall and Wellner (1980) for the computation of confidence bands for censored data.

Hall and Wellner define the following terms:

$$C_n(t) = n\left[\frac{\text{S.}E.[\hat{S}(t)]}{\hat{S}(t)}\right]^2$$
$$K_n(t) = \frac{C_n(t)}{1+C_n(t)}$$
$$\overline{K}_n(t) = 1 - K_n(t) = [1+C_n(t)]^{-1}.$$

They then prove that

$$\lim_{n\to\infty} P\{\hat{S}_n(t) - \frac{d_{n,(1-\alpha)}}{\sqrt{n}}\left[\frac{\hat{S}_n(t)}{\overline{K}_n(t)}\right] < S(t)$$
$$< \hat{S}_n(t) + \frac{d_{n,(1-\alpha)}}{\sqrt{n}}\left[\frac{\hat{S}_n(t)}{\overline{K}_n(t)}\right]\} = 1-\alpha$$

for all $t \leq t_{max}$.

For the validity of the theory, t_{max} should be set equal to next to largest rather than largest observed survival time. Hall and Wellner acknowledged their formula to be somewhat conservative and provide a table of smaller values of $d\alpha$, which may be used when the term $1-\overline{k}_n(t_{max})$ occurs unless the estimated survival rate at t_{max} is still fairly high, say, greater than 0.4.

Table 5.7

Critical values of $d_{(1-\alpha)}$ to be used for Hall-Wellner Confidence Bands when $[1-\overline{k}_n(t_{max})] < 0.75$.

	$1-\overline{k}_n(t_{max})$				
$1-\alpha$	0.25	0.40	0.50	0.60	0.75
0.99	1.256	1.47	1.552	1.60	1.626
0.95	1.014	1.198	1.273	1.321	1.354
0.90	0.894	1.062	1.133	1.181	1.217

The following illustration uses critical values for the Hall-Wellner confidence band from the book by Klein and Moeschberger (2003).

Illustration 5.5 (Harris and Albert (1991)): Fortier et al. (1986) have investigated the effects of varying dosages of preoperative radiation therapy in rectal cancer. The following table shows the survival times in months of 35 patients receiving the dosage $\geq$ 5000 rad.

Table 5.8
Survival times of patients with rectal cancer

Sr. No. of Patient	Survival time (months)	Sr. No. of patient	Survival time (months)	Sr. No. of patient	Survival time (months)
1	9	13	27	25	38^+
2	12	14	29^+	26	51^+
3	13^+	15	30^+	27	54^+
4	14^+	16	32^+	28	57
5	14^+	17	33^+	29	60^+
6	16	18	33^+	30	67
7	18^+	19	35	31	70
8	19	20	35	32	87^+
9	23^+	21	35^+	33	89^+
10	24^+	22	35^+	34	98^+
11	25^+	23	35^+	35	120^+
12	26^+	24	36		

Table 5.9
Kaplan-Meier estimates of $(S(t))$

t_j	$\hat{S}(t)$
9	0.971
12	0.943
16	0.911
19	0.879

Table 5.9 Cont'd

t_j	$\hat{S}(t)$
27	0.841
35	0.742
36	0.680
57	0.595
67	0.496

Figure 5.4 shows the survival function and the confidence bands.

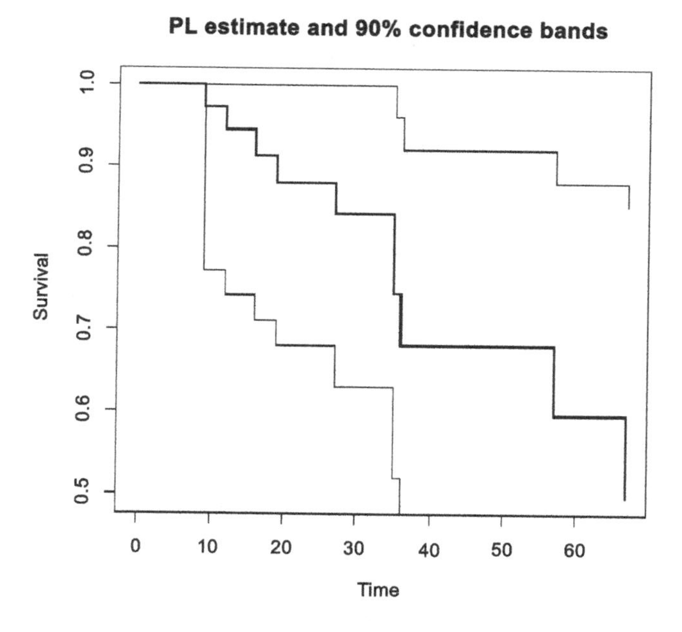

Figure 5.4

For R-commands see the appendix of this chapter.

Redistribution to the Right Algorithm

This is another intuitive method of computing the estimator of survival function which is introduced by Efron. We illustrate this with the Leukemia (AML) example by considering maintained group. Plot the 11 survival times.

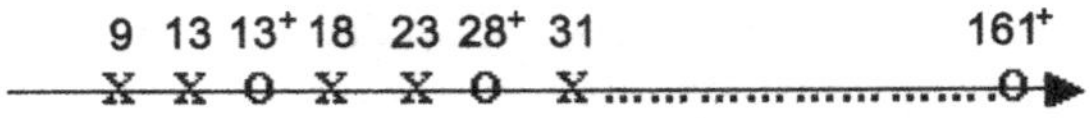

9 13 13⁺ 18 23 28⁺ 31 161⁺

X X O X X O X .... O

The ordinary estimate of $S(t)$ assuming no censoring puts mass $\frac{1}{n} = \frac{1}{11}$ at each observed time. Consider the first censored time viz. 13^+. Since death did not occur at 13 but to the right of it, it is reasonable to distribute the mass equally among the points which are to the right of it. Therefore, add $\frac{1}{8} \times \frac{1}{11}$ to the mass of observations 18, 23, ..., 161^+ which are to the right of 13^+. Consider, next censored observation viz. 28^+. Now this observation has mass $\frac{1}{11} + \frac{1}{8} \times \frac{1}{11}$. Redistribute it equally among the points 31, ..., 161^+ . Treat the other censored times similarly.

The following table shows the computations and final estimates of survival function using this algorithm.

Table 5.10

Survival function estimate by using redistribution to the right algorithm

Sr.No.	$T_{(j)}$	Mass at the start	Mass after redistribution			$\hat{S}(T_{(j)})$
			No. 1	No. 2	No. 3	
1	9	0.0909	0.0909	0.0909	0.0909	0.9091
2	13	0.0909	0.0909	0.0909	0.0909	0.8265
3	13^+	0.0909	0	0	0	-
4	18	0.0909	0.1023	0.1023	0.1023	0.7419
5	23	0.0909	0.1023	0.1023	0.1023	0.6660
6	28^+	0.0909	0.1023	0	0	-
7	31	0.0909	0.1023	0.1228	0.1228	0.5843
8	34	0.0909	0.1023	0.1228	0.1228	0.5125
9	45^+	0.0909	0.1023	0.1228	0	-
10	48	0.0909	0.1023	0.1228	0.1842	0.4181
11	161^+	0.0909	0.1023	0.1228	0.1842	-

Generalized Maximum Likelihood Estimator

In the usual set up, we assume that our observation vector say $\underline{X}$ has a probability measure P_θ that satisfies

$$dP_\theta(\underline{x}) = f_\theta(\underline{x})d\mu(\underline{x})$$

where $\mu(\underline{x})$ is a dominating measure for the class of measures $\{P_\theta\}$. Getting the maximum likelihood estimator of θ involves maximising the likelihood

$$L(\theta) = f_\theta(\underline{x}) = dP_\theta(\underline{x})/d\mu(\underline{x}).$$

In the non-parametric set up, we assume that our observation vector has a probability measure P_F that depends on the unknown distribution function F. The class $\{P_F\}$ has no dominating measure so we need a more general definition of maximum likelihood.

Kiefer and Wolfowitz (1956) suggest the following definition. Let $\mathcal{P} = \{P\}$ be a class of probability measures. For the elements P_1 and P_2 in $\mathcal{P}$, define

$$f(\underline{x}; P_1, P_2) = \frac{dP_1(\underline{x})}{d(P_1 + P_2)},$$

the Radon-Nikodym derivative of P_1 with respect to $P_1 + P_2$. Define the probability measure $\hat{P}$ to be generalised maximum likelihood estimator (GMLE) if

$$f(\underline{x}; \hat{P}, P) \geq f(x; P, \hat{P}) \tag{5.3.1}$$

for any element $P \in \mathcal{P}$.

Theorem 5.3.1: The Kaplan-Meier estimator is the GMLE of survival function.

Proof. For convenience assume no ties.

If a probability measure $\hat{P}$ gives positive probability to $\underline{X}$, then $f(\underline{x}; P, \hat{P}) = 0$ unless P also gives positive probability to $\underline{\mathrm{X}}$. Thus, to check (5.3.1) it is enough to consider those probability measures P for which $P\{\underline{X}\} > 0$ and in this case (5.3.1) reduces to

$$\hat{P}\{\underline{X}\} \geq P\{\underline{X}\}. \tag{5.3.2}$$

Since $\hat{S}$ puts positive mass on the point

$$\underline{X} = \{(T_1, \delta_1), (T_2, \delta_2), ..., (T_n, \delta_n)\},$$

we need only to consider probability measures P which put positive mass on this point in $\mathcal{R}^n$ and show that $\hat{S}$ maximises

$$P\{[(T_1, \delta_1), (T_2, \delta_2), ..., (T_n, \delta_n)]\}.$$

For any such P,

$$L = P\{(T_1, \delta_1), (T_2, \delta_2), ..., (T_n, \delta_n)]\}$$
$$= \prod_{i=1}^{n} \{P[T = T_{(i)}]\}^{\delta_{(i)}} \{P[T > T_{(i)}]\}^{1-\delta_{(i)}}.$$

Let P be a measure which assigns the probability p_i to the half open interval $[T_{(i)}, T_{(i+1)})$, where $T_{(n+1)} = \infty$. For fixed $p_1, p_2, ..., p_n$ the likelihood L is maximised by setting $P[T = T_{(i)}] = p_i$ if $\delta_{(i)} = 1$ where p_i is the probability of failure. If $\delta_{(i)} = 0$ then L is maximised by setting $P[T_{(j)} < T < T_{(j+1)}] = p_j, j = i, i+1, ..., n$. Thus, for fixed $p_1, p_2, ..., p_n$ the maximum value of L is

$$\prod_{i=1}^{n} p_i^{\delta_{(i)}} (\sum_{j=i}^{n} p_j)^{1-\delta_{(i)}}. \tag{5.3.3}$$

To show that PL estimator is GMLE is equivalent to show that the maximum of (5.3.3) is attained for

$$p_i = \frac{\delta_{(i)}}{(n-i+1)} \prod_{j=1}^{i-1} (1 - \frac{\delta_{(i)}}{n-j+1}).$$

For, let

$$\lambda_i = \frac{p_i}{\sum_{j=i}^{n} p_j}, \qquad i = 1, 2, ..., n. \tag{5.3.4}$$

Then since

$$(1 - \lambda_i) = \frac{\sum_{j=i+1}^{n} p_j}{\sum_{j=i}^{n} p_j} \quad \text{and} \quad \sum_{j=1}^{n} p_j = 1,$$

we have by mathematical induction

$$\sum_{j=i}^{n} p_j = \prod_{j=1}^{i-1} (1 - \lambda_j). \tag{5.3.5}$$

Therefore by using (5.3.5) and (5.3.4),

$$p_i = \lambda_i \prod_{j=1}^{i-1} (1 - \lambda_j),$$

$$\prod_{i=1}^{n}[p_i^{\delta_{(i)}}(\sum_{j=i}^{n}p_j)^{1-\delta_{(i)}}]$$
$$=\prod_{i=1}^{n}\{\lambda_i^{\delta_{(i)}}[\prod_{j=1}^{(i-1)}(1-\lambda_j)]^{\delta_{(i)}}\}$$
$$\times[\prod_{j=1}^{(i-1)}(1-\lambda_j)]^{1-\delta_{(i)}}\prod_{i=1}^{n-1}\lambda_i^{\delta_{(i)}}(1-\lambda_i)^{(n-i)} \tag{5.3.6}$$

(since $\lambda_n = 1$).

To maximise (5.3.6) we differentiate $\log[\lambda_i^{\delta_{(i)}}(1-\lambda_i)^{n-i}] = \delta_{(i)}\log\lambda_i + (n-i)\log(1-\lambda_i)$ and equate the derivative to 0. Solving for λ_i, then yields $\lambda_i = \frac{\delta_{(i)}}{n-i+\delta_{(i)}}$.

Verifying that

$$\frac{\delta^2}{\delta^2\lambda_i}\log[\lambda_i^{\delta_{(i)}}(1-\lambda_i)^{n-i}] < 0.$$

We conclude that

$$\hat{\lambda}_i = \frac{\delta_{(i)}}{n-i+\delta_{(i)}} = \frac{\delta_{(i)}}{n-i+1} \tag{5.3.7}$$

and

$$\hat{p}_i = \frac{\delta_{(i)}}{(n-i+1)}\prod_{j=1}^{(i-1)}(1-\frac{\delta_{(j)}}{(n-j+1)})$$

are the corresponding maximum likelihood estimators. Putting these together we get $\hat{S}$, the Kaplan-Meier estimator. The argument for data with ties is on similar lines.

Properties of Kaplan-Meier Estimator

(i) PL estimator is consistent.

(ii) Asymptotic Normality: If F, the lifetime distribution and G, the censoring distribution are continuous on $[0, T]$ and $F(T) < 1$, then

$$Z_n(t) = \sqrt{n}[\hat{S}(t) - S(t)] \xrightarrow{W} Z(t) \text{ as } n \to \infty \text{ for each } t,$$

where $\{Z(t)\}$ is a Gaussian process with moments

$$E(Z(t)) = 0.$$
$$Cov[Z(t_1), Z(t_2)] = S(t_1)S(t_2) \times \int_0^{t_1 \Lambda t_2}\frac{dF_u(u)}{(1-F(u))[1-H(u)]},$$

where

$$\begin{aligned} F_u(t) &= P[T \le t, \delta = 1] \\ &= \int_0^t (1 - G(u))dF(u) \\ \& \ (1 - H(u)) &= [1 - F(u)][1 - G(u)]. \end{aligned}$$

Hence $f(Z_n)$ converges in distribution to $f(Z)$ for any function f continuous in the supnorm. As a particular case of the result

$$\hat{S}(t) \overset{a}{\sim} N(S(t), \frac{S^2(t)}{n} \int_0^t \frac{dF_u(u)}{[1 - H(u)]^2}).$$

Since

$$F_u(t) = P[T \le t, \delta = 1]$$

and

$$H(t) = P[T \le t].$$

$$\begin{aligned} 1 - \hat{H}(T_{(i)}) &= 1 - \frac{i}{n} = \frac{(n-i)}{n} \\ 1 - \hat{H}(T_{(i)}^-) &= \frac{(n-i+1)}{n}. \end{aligned}$$

Replacement $(1 - H(u))^2$ by $(1 - H(u))(1 - H(u^-))$ in the asymptotic variance and substitution of the estimates gives the following estimate of asymptotic variance of $\hat{S}(t)$.

$$\begin{aligned} A\widehat{Var}\hat{S}(t) &= \frac{(\hat{S}(t))^2}{n} \sum_{T_{(i)} \le t} \frac{\frac{\delta_{(i)}/n}{(n-i)(n-i+1)}}{n^2} \\ &= (\hat{S}(t))^2 \sum_{T_{(i)} \le t} \frac{\delta_{(i)}}{(n-i)(n-i+1)} \end{aligned}$$

which is precisely Greenwood's formula. (Breslow and Crowley (1974)).

Hazard Function Estimators

The hazard rate is

$$r(t) = \frac{f(t)}{\overline{F}(t)} = \frac{f(t)}{S(t)}.$$

Estimating $r(t)$ is essentially equivalent to the difficult problem of estimating a density. An easier problem is estimating the *cumulative hazard function* ($\Lambda(t)$) given by $\Lambda(t) = \int_0^t r(u)du$. Then

$$S(t) = e^{-\Lambda(t)}.$$

For the sake of simpler notation, assume no ties, Nelson (1969) estimates $\Lambda(t)$ by

$$\hat{\Lambda}(t) = \hat{\Lambda}_2(t) = \sum_{T_{(i)} \leq t} \frac{\delta_{(i)}}{(n-i+1)}$$

and Peterson (1977) proposes

$$\hat{\Lambda}_1(t) = \sum_{T_{(i)} \leq t} -\log(1 - \frac{\delta_{(i)}}{n-i+1}).$$

Since $\log(1-x) = -x$, for small values of x, the two estimators are close for small values of t. Peterson's estimator corresponds to the PL estimator of the survival function

$$\begin{aligned} \hat{S}_1(t) = e^{-\hat{\Lambda}(t)} &= \prod_{T_{(i)} \leq t} (1 - \frac{\delta_{(i)}}{n-i+1}) \\ &= \hat{S}(t). \end{aligned}$$

While Nelson's estimator corresponds to a different estimator of the survival function

$$\hat{S}_2(t) = e^{-\hat{\Lambda}_2(t)}.$$

Fleming and Harrington (1979) have recommended $\hat{S}_2(t)$ as an alternative estimator of survival function and have shown it to have slightly smaller mean square error in some situations.
Estimation of the Mean of the Distribution

$$\begin{aligned} \mu = &\text{ Mean of the lifetime distribution} \\ = &\int_0^\infty x dF(x) \\ = &\int_0^\infty S(t)dt. \end{aligned}$$

For complete data,

$$\begin{aligned}\hat{\mu} &= \int_0^{\infty} [\overline{F}_n(x)]dx = \int_0^{\infty} xdF_n(x)\\ &= \frac{1}{n}\sum_{i=1}^{n} X_i = \overline{X}.\end{aligned}$$

For censored data
$\hat{\mu} = \int_0^{\infty} \hat{S}(t)dt =$ Area under graph of $\hat{S}(t)$ (*)
$= \int_0^{\infty} xd\hat{F}(x)$ (Integral reduces to the sum).

If the last observation is censored, then $\hat{S}(t)$ does not approach zero as $t \to \infty$ so the integral on the R.H.S. of (*) is inifinite. In what follows, we discuss the possible solutions.

(1) *Redefinition of the last observation*: Change $\delta_{(n)} = 0$ to $\delta_{(n)} = 1$. We illustrate with the AML data for the maintained group.

$$\begin{aligned}\hat{\mu} &= 9[1-0.91] + 13[0.91-0.82]\\ &\quad +18[0.82-0.72] + 23[0.72-0.61]\\ &\quad +31[0.61-0.49] + 34[0.49-0.37]\\ &\quad +48[0.37-0.18] + 161[0.18]\\ &= 52.635.\end{aligned}$$

In the above computations the survival probabilities: 0.91, 0.82, 0.72, 0.61, 0.49, 0.37, 0.18 are obtained by using R-commands similar to those for Illustration 5.3.

Observe that the last observation has heavy weight.

(2) *Restricted Mean*: For fixed T_0 define a mean over $(0, T_0]$ and estimate it by

$$\begin{aligned}\hat{\mu} &= \int_0^{T_0} xd\hat{F}(x)\\ &= 23.011\end{aligned}$$

with $T_0 = 48$ for the same problem.

(3) *Variable Upper Limit*: Estimate

$$\mu = \int_0^{\infty} S(t)dt \text{ by } \hat{\mu} = \int_0^{s_n} \hat{S}(t)dt,$$

where $\{s_n\}$ is a sequence of numbers converging monotonically to ∞. Unfortunately the proper choice of s_n depends on F, G and there are no good guidelines for the use of this estimator in practice.
Remark. In the above illustration we have used $\hat{\mu} = \int_0^\infty x d\hat{F}(x)$. However, one can use

$$\int_0^\infty \hat{S}(t)dt = \hat{\mu}.$$

In this case, one finds area under the graph of the survival function. We shall illustrate the procedure with AMI data for the non-maintained group.

Table 5.11
Estimation of the mean

Time Interval $[\ell_1, \ell_2)$	$\hat{S}(t)$
0 - 5	1
5 - 8	0.8333
8 - 12	0.6667
12 - 23	0.5834
23 - 27	0.4862
27 - 30	0.3884
30 - 33	0.2917
33 - 43	0.1945
43 - 45	0.0972

$$\begin{aligned}
\hat{\mu} &= \int_0^{45} \hat{S}(t)dt \\
&= 1 \times 5 + 0.8333 \times 3 + 0.6667 \times 4 \\
&\quad +0.5834 \times 11 + 0.4862 \times 4 + 0.3884 \times 3 \\
&\quad +0.2917 \times 3 + 0.1945 \times 10 + 0.972 \times 2 \\
&= 22.71.
\end{aligned}$$

For R-commands see the appendix of this chapter.

Exercises

Exercise 5.1: Consider the data of patients with ovary cancer diagnosed in Connecticut from 1935 to 1944 (Cutler et al. 1960b). The intervals, number of individuals at risk, number of deaths and number of censored individuals are given in the following table. Estimate the survival function and obtain an estimate of standard error of estimator of the survival function for each interval using Greenwood's formula.

Table 5.12

Serial Number	Class interval	Number at risk	Number of deaths	Number censored
1	0 - 8	21	4	1
2	8 - 12	16	1	3
3	12 - 18	12	2	1
4	18 - 24	9	2	2
5	24 - 36	5	0	5

Exercise 5.2: The following table shows the times to recurrence of brain metastasis for a sample of 23 patients treated with radiotherapy. Censored observations are shown with + sign. Prepare a plot of Kaplan-Meier estimate of survival function with pointwise 95% confidence intervals.

Table 5.13

Patient No.	Time (Weeks)	Patient No.	Time (Weeks)
1	2+	13	14
2	2+	14	14+
3	2+	15	18+
4	3	16	19+
5	4	17	20
6	5	18	22
7	5+	19	22+
8	6	20	31+
9	7	21	33
10	8	22	39
11	9+	23	195+
12	10		

Exercise 5.3: We wish to compare the times to recurrence of brain metastasis for patients treated with radiotherapy alone (group 1) and for those undergoing surgical removal of the tumor and subsequent radiotherapy (group 2). Survival times for both the groups are presented in the following table.

Radiotherapy Alone		Surgery/Radiotherapy	
Patient number	Recurrence (Weeks)	Patient number	Recurrence (Weeks)
1	8	1	29+
2	9+	2	32+
3	10	3	34+
4	14	4	34+
5	14+	5	37
6	18+	6	37+
7	19+	7	42+
8	20	8	51
9	22	9	57
10	22+	10	59

Carry out graphical test for the hypothesis that group 2 has better survival experience than group 1.

Exercise 5.4: The following data are survival times in weeks of patients with lymphocytic non-Hodgkins (see Dinse (1982) and Kimber (1990)). The asterisks denote censored survival times. Plot group-wise survival curves to compare the two populations.

Group I: Asymptomatic

50, 58, 96, 139, 152, 159, 189, 225, 239, 242, 257, 262, 292, 294, 300*, 301, 306*, 329*, 342*, 346*, 349*, 354*, 359, 360*, 365*, 378*, 381*, 388*, 281, 362*.

Group II: Symptomatic

49, 58, 75, 110, 112, 132, 151, 276.

Exercise 5.5: Show that in the absence of censoring, product - limit estimator of survival function reduces to the empirical survival function. Show also that Greenwood's formula for variance of product limit estimator of survival function reduces to the usual variance estimator of the empirical survival function.

Exercise 5.6: An experiment was conducted to determine the effect of a

drug on leukemia remission times. The following are the survival times of 10 patients:
6, 6, 6, 6+, 7, 9+,10, 10+, 11+, 13.
Use redistribution to the right algorithm to estimate the survival probabilities.
Exercise 5.7 (Nair (1984)): The data in the following table are the failure times (measured in millions operations) of 40 randomly chosen mechanical switches. They were tested in a facility with 40 test positions. Three of the test positions became available much later than the others, so the three switches tested at this position were still operating at the termination of the test. The corresponding censored observations are indicated by code 0 in the table. There were two possible modes of failures-essentially two different springs- for the switches. When a switch failed its mode of failure was noted. These modes are indicated by codes A(1) and B(0) in the table under the heading status. The two springs were identical in construction but subjected to different stress levels, so the life distributions of the two failure modes are likely to be different. We shall focus on the estimation of the survival for failure mode A. This mode is chosen because there is lot of censoring in the upper tail of the distribution of B. Every occurrence time for failure B is considered as censoring time for mode A.

Table 5.14

Time	Status	Time	Status
1.151	0	2.119	0
1.17	0	2.135	1
1.248	0	2.197	1
1.331	0	2.199	0
1.381	0	2.227	1
1.499	1	2.25	0
1.508	0	2.254	1
1.534	0	2.261	0
1.577	0	2.349	0

Table 5.14 Cont'd

Time	Status	Time	Status
1.584	0	2.369	1
1.667	1	2.547	1
1.695	1	2.548	1
1.71	1	2.738	0
1.955	0	2.794	1
1.965	1	2.883	0
2.012	0	2.883	0
2.051	0	2.91	1
2.076	0	3.015	1
2.109	1	3.017	1
2.118	0	3.793	0

Construct a plot of product limit estimate of reliability function with 95% confidence bands by Hall and Wellner.

Exercise 5.8: The following remissions are observed from 10 patients with solid tumours. Six patients relapse at 3.0, 6.5, 6.5, 10, 12 and 15 months; 1 patient is lost to follow-up at 8.4 months; and three patients are still in remission at the end of study after 4.0, 5.7, and 10 months. Plot survival function and obtain from the plot median survival time. Calculate an estimate of mean survival time.

Exercise 5.9: A laboratory investigator interested in the relationship between diet and the development of tumours divided rats in three groups and fed them with low-fat, saturated and unsaturated diats. (King et al. (1979)). The following survival times are subset of survival times of rats with low fat diet: 140, 177, 50, 65, 86, 153, 181, 191, 77, 84, 87, 56, 66, 73, 119 and 140+ (+ indicates censored observation). Obtain an estimate of mean survival time.

Appendix

Non-parametric estimator of survival function

R-commands to plot empirical survival function.

```
> library(survival); # Attach the package survival;
> time<- c(17.88,28.92,33.0,41.52,42.12,45.60,48.48,51.84,
```

```
  51.96,54.12,55.56,67.80, 68.64,68.64,68.88,84.12,93.12,
  98.64,105.12,105.84,127.92,128.04,173.4);
> # Create a vector of survival times;
> length(time);
[1] 23
> status<-c(rep(1,23));
> # Create a vector which indicates whether the survival
> # time is censored or complete;
> # 1 indicates complete and zero indicates censored;
> length(status);
[1] 23
> ball<-data.frame(time,status);
> # Create a data set of the vectors time and status;
> attach(ball);
The following objects are masked _by_ .GlobalEnv:
    status, time

> km.ball<-survfit(Surv(time,status)~1) #Creates survival object;
> plot(km.ball,conf.int=F,xlab="time",ylab="survival function",
  main="Non-parametric estimator of survival function",cex=.6);
> # Plot empirical survival curve;
```

The second argument to plot function indicates that pointwise confidence interal is not required.

Acturial Estimator of Reliability Function (Illustration 5.2)

```
> n<-c(200,195,184,167,157,147,126,114,105,101);
> length(n);
[1] 10
> r<-c(0,1,5,2,0,6,3,1,0,1);
> nprime<-(n-(0.5*r));
> nprime;
[1] 200.0 194.5 181.5 166.0 157.0 144.0 124.5 113.5 105.0 100.5
> d<-c(5,10,12,8,10,15,9,8,4,3);
> q<-d/nprime;
> q<-round(q,3);
> p<-1-q;
> p;
```

```
[1] 0.975 0.949 0.934 0.952 0.936 0.896 0.928 0.930 0.962 0.970
> survp<-cumprod(p);
> survp<-round(survp,3);
> survp;
[1] 0.975 0.925 0.864 0.823 0.770 0.690 0.640 0.595 0.573 0.556
```

Computation of standard error of actuarial estimator at $t = 5$:

```
> t <-5; st <- 0.770; nprime <-c(200,194.5,181.5,166,157);
> d <- c(5, 10, 12, 8, 10);
> var.st <-st ^ 2*sum(d/(nprime*(nprime-d)));
> var.st;
[1] 0.0009102475
> se <- sqrt(var.st);
> se;
[1] 0.03017031
```

Kaplan-Meier estimate for failures of vanes (Illustration 5.3)

```
> time<-c(142,149,320,345,560,805,1130,1720,2480,4210,5230,6890);
>  length(time);
[1] 12
> status<-c(1,1,1,0,1,1,0,1,0,0,1,1);
> length(status);
[1] 12
> ctrl<-data.frame(time,status);

> attach(ctrl);
The following objects are masked _by_ .GlobalEnv:
    status, time
The following objects are masked from ball:
    status, time

> library(survival);
> km.ctrl<-survfit(Surv(time,status==1)~ 1);
> summary(km.ctrl);
Call: survfit(formula = Surv(time, status == 1) ~ 1)

time n.risk n.event survival std.err lower 95% CI upper 95%CI
 142     12       1    0.917  0.0798       0.7729       1.000
```

```
 149    11      1    0.833  0.1076        0.6470       1.000
 320    10      1    0.750  0.1250        0.5410       1.000
 560     8      1    0.656  0.1402        0.4318       0.997
 805     7      1    0.562  0.1482        0.3356       0.943
1720     5      1    0.450  0.1555        0.2286       0.886
5230     2      1    0.225  0.1771        0.0481       1.000
6890     1      1    0.000     NaN            NA          NA
```

Groupwise survival curves (Illustration 5.4)

```
> time<-c(5, 5,8,8,12,16,23,27,30, 33, 43, 45, 9, 13, 13,
  18, 23, 28, 31, 34, 45, 48, 161);
> length(time);
[1] 23
> status<-c(rep(1,5),0,rep(1,6),1,1,0,1,1,0,1,1,0,1,0);

> gr<-c(rep(1,12),rep(2,11));
> ctc<-data.frame(time,status,gr);
> ctc;
   time status gr
1     5      1  1
2     5      1  1
3     8      1  1
4     8      1  1
5    12      1  1
6    16      0  1
7    23      1  1
8    27      1  1
9    30      1  1
10   33      1  1
11   43      1  1
12   45      1  1
13    9      1  2
14   13      1  2
15   13      0  2
16   18      1  2
17   23      1  2
18   28      0  2
19   31      1  2
```

```
20   34      1  2
21   45      0  2
22   48      1  2
23  161      0  2

> attach(ctc);
The following objects are masked _by_ .GlobalEnv:
    gr, status, time
The following objects are masked from ctrl:
    status, time
The following objects are masked from ball:
    status, time
The following object is masked from larynx:
    time

> library(survival);
> surv.bygr<-survfit(Surv(time,status==1)~ gr);

> plot(surv.bygr,conf.int=F,xlab="time",
  ylab="survival function",
  main="Survival curves for the two groups",cex=.6);
> text(locator(1),"Upper curve represents Maintained group",
  cex=.8);
> text(locator(1),"and",cex=.8);
> text(locator(1),
  "Lower curve represents Non-maintained group",cex=.8);

> l1<-c(0,5,8,12,23,27,30,33,43);
> l2<-c(5,8,12,23,27,30,33,43,45);
> w<-l2 - l1;
> sp<-c(1,0.8333,0.6667,0.5834,0.4862,0.3884,0.2917,
  0.1945,0.0972);
> estmu<-sum(w*sp);
> estmu;
[1] 22.7086
```

Hall and Wellner confidence bands (Illustration 5.5):

```
> library (survival);
```

```
> time <- c(9, 12, 13, 14, 14, 16, 18, 19, 23, 24, 25, 26,
  27, 29, 30, 32, 33, 33, 35, 35, 35, 35, 35, 36, 38, 51,
  54, 57, 60, 67, 70, 87, 89, 98, 120);

> length (time);
[1] 35
> status <- c(1, 1, 0, 0, 0, 1, 0, 1, 0, 0, 0, 0, 1, 0, 0,
 0, 0, 0, 1, 1, 0, 0, 0, 1, 0, 0, 0, 1, 0, 1, 1, 0, 0, 0,0);
> length (status);
[1] 35

> d<- data.frame(time, status);
> fit0<-survfit(Surv(time,status)~1,conf.int=0.90,data=d);
> summary (fit0);

Call:survfit(formula=Surv(time,status)~1,data=d,conf.int=0.9)

time n.risk n.event survival std.err lower 90% CI upper 90%CI
   9     35       1    0.971  0.0282        0.926       1.000
  12     34       1    0.943  0.0392        0.880       1.000
  16     30       1    0.911  0.0489        0.834       0.996
  19     28       1    0.879  0.0570        0.790       0.978
  27     23       1    0.841  0.0661        0.739       0.957
  35     17       2    0.742  0.0878        0.610       0.901
  36     12       1    0.680  0.0999        0.534       0.866
  57      8       1    0.595  0.1182        0.429       0.825
  67      6       1    0.496  0.1338        0.318       0.773
  70      5       1    0.397  0.1390        0.223       0.706

> library (km.ci);
> nfit0 <- km.ci(fit0,method = "hall-wellner");
> par(font = 2, font.axis = 2, font.lab=2);
> plot(nfit0,main="PL estimate & 90% cofnidence bands",lwd=2);
```

Chapter 6

Tests for Exponentiality

6.1 Introduction

The exponential distribution plays an important role in reliability and life-time modelling, just as the normal distribution in classical statistics. For this distribution, explicit and simple forms of survival function, density and hazard are available. It is technically convenient for drawing inferences even in the presence of censoring. Furthermore it is the only distribution with the memoryless (no ageing) property and therefore is often used to model the lifetimes of electronic and other non-ageing components. However, exponential distribution should be used judiciously since its no ageing property actually restricts its applicability. Many mechanical components undergo wear (e.g. bearings) or fatigue (e.g. structural components) whereas certain electronic components undergo reliability growth. These are the reasons why testing for exponentiality is important and why there are many tests of exponentiality. However, out of the several tests for exponentiality we shall only study the three tests: (i) Hollander and Proschan's test (1972), (ii) certain tests based on sample spacings (Hollander and Proschan (1975)), Klefsjo (1983) and (iii) Deshpande's class of tests (1983).

For these tests of exponentiality we need at least a working knowledge of a powerful technique of non-parametric inference known as U-statistics. Hence before discussing these tests we shall get acquainted with the use of this tool.

6.2 U-Statistics

Definition 6.2.1: Let $X_1, X_2, \cdots, X_n$ be a random sample from the distribution $F \in I\!\!F$. A parameter γ is said to be estimable of degree r for the family of distributions $\mathbb{F}$ if r is the smallest sample size for which there exists a function $h^*(x_1, x_2, \cdots, x_r)$ such that

$$E_F[h^*(X_1, X_2, \cdots, X_r)] = \gamma \tag{6.2.1}$$

for every $F \in \mathbb{F}$.

The function $h^*(\cdot)$ in (6.2.1) is known as the kernel for the parameter γ.

It may be noted that for any kernel $h^*(X_1, X_2, \cdots, X_r)$ we can always create one that is symmetric in its arguments by using

$$h(X_1, X_2, \cdots, X_r) = \frac{1}{r!} \sum_A h^*(X_{\alpha_1}, \cdots, X_{\alpha_r})$$

where $\alpha_1, \alpha_2, \cdots, \alpha_r$ is a permutation of the numbers $1, 2, \cdots, r$ and A is the set of all permutation $(\alpha_1, \cdots, \alpha_r)$ of the integers $1, 2, \cdots, r$.

It is easy to see that h is also unbiased for γ. Hence without loss of generality we shall assume that the kernel h is symmetric.

We have a random sample of size $n (n \geq r)$ from the distribution $F \in \mathbb{F}$. Naturally, we want to use all the n observations to construct an unbiased estimator of γ.

A U-statistic for the estimable function γ is constructed with the symmetric kernel $h(.)$ by forming

$$U(X_1, \cdots, X_n) = \frac{1}{\binom{n}{r}} \sum_{\underline{\beta} \in \mathcal{B}} h(X_{\beta_1}, \cdots, X_{\beta_r}),$$

where $\underline{\beta} = (\beta_1, \cdots, \beta_r)$ is a combination of r integers from $(1, 2, \cdots, n)$ and $\mathcal{B}$ is the set of all such combinations.

It can be shown that the U-statistic thus constructed is the unique MVUE (Minimum Variance Unbiased Estimator) of γ.

Illustration 6.1

(1) Let $\mathbb{F}$ denote the class of all distributions with finite first moment γ

$$E_F(X_1) = \gamma \quad \forall \quad F \in \mathbb{F}.$$

Thus, mean is an estimable parameter of degree 1. Here $h(x) = x$ is trivially symmetric. The U-statistic estimator of mean is

$$U(X_1, X_2, \cdots, X_n) = \frac{1}{n} \sum_{i=1}^{n} X_i = \overline{X}.$$

(2) Let $I\!F$ denote the collection of all distributions with finite variance γ.

$$E_F[X_1^2 - X_1X_2] = \gamma \quad \forall \; F \in I\!F.$$

Thus variance is estimable and of degree 2. It may be noted here that to estimate variance (or any other measure of variability) at least two observations are essential. The associated symmetric kernel is

$$\begin{aligned} h(x_1, x_2) &= \frac{1}{2}[(X_1^2 - X_1X_2) + (X_2^2 - X_1X_2)] \\ &= \frac{1}{2}(X_1 - X_2)^2. \end{aligned} \tag{6.2.2}$$

The U-statistic uses the symmetric kernel (6.2.2) to form

$$\begin{aligned} U(X_1, \cdots, X_n) &= \frac{1}{\binom{n}{r}} \frac{1}{2}[(n-1)\sum_1^n X_i^2 - 2\sum\sum_{i<j} X_iX_j] \\ &= \frac{1}{(n-1)}[\sum_1^n X_i^2 - n\overline{X}^2] \\ &= s^2. \end{aligned}$$

Variance of the U-statistic.

For a symmetric kernel $h(.)$ consider the random functions $h(X_1, \cdots, X_c, X_{c+1}, \cdots, X_r)$ and $h(X_1, \cdots, X_c, X_{r+1}, \cdots, X_{2r-c})$ having exactly c variables in common. The covariance between these two random variables is given by

$$\begin{aligned} \xi_c &= Cov[h(X_1, \cdots, X_c, X_{c+1}, \cdots, X_r), h(X_1, \cdots, X_c, X_{r+1}, \cdots, X_{2r-c})]. \\ &= E[h(X_1, \cdots, X_c, X_{c+1}, \cdots, X_r)h(X_1, \cdots, X_c, X_{r+1}, \cdots X_{2r-c})] - \gamma^2. \end{aligned} \tag{6.2.3}$$

Further,

$$\xi_c = Cov[h(X_{\beta_1}, \cdots, X_{\beta_r}), h(X_{\beta_1'}, \cdots, X_{\beta_r'})],$$

where $(\beta_1, \cdots, \beta_r)'$ and $(\beta_1', \cdots, \beta_r')'$ are subsets of the integers $\{1, 2, \cdots, n\}$ having exactly c integers (out of r) in common. It may be noted that if $c = 0$ then the kernel functions based on β and β' are independent. Hence

$$\xi_0 = 0. \tag{6.2.4}$$

Now the variance of the U-statistic is

$$Var(U) = E[\{\frac{1}{\binom{n}{r}} \sum_{\underline{\beta} \in \mathcal{B}} h(X_{\beta_1}, \cdots, X_{\beta_r}) - \gamma]\}^2]$$

$$= \frac{1}{\binom{n}{r}^2} \sum_{\beta} \sum_{\beta'} Cov[h(X_{\beta_1}, \cdots, X_{\beta_r}), h(X_{\beta'_1}, \cdots, X_{\beta'_r})]. \quad (6.2.5)$$

All the terms in (6.2.5) for which $\underline{\beta}$ and $\underline{\beta}'$ have exactly c integers in common have the same covariance, say, ξ_c. The number of such terms is $\binom{n}{r}\binom{r}{c}\binom{n-r}{r-c}$. It follows that

$$Var(U) = \frac{1}{\binom{n}{r}} \sum_{c=1}^{r} \binom{r}{c}\binom{n-r}{r-c} \xi_c \quad \text{since } \xi_0 = 0. \qquad (6.2.6)$$

Illustration 6.2

(1) Consider the population parameter

$$\gamma = P[X_1 + X_2 > 0],$$

where X_1, X_2 are independent observations from F. It follows that

$$\psi(X_1, X_2) = \begin{cases} 1 & \text{if } X_1 + X_2 > 0 \\ 0 & \text{otherwise} \end{cases}$$

is a symmetric kernel of degree 2. So the corresponding U-statistic is

$$U(X_1, \cdots, X_n) = \frac{1}{\binom{n}{2}} \sum_{i<j}\sum \psi(X_i, X_j)$$

$$= \frac{1}{\binom{n}{2}} \{\# \text{ of pairs } (X_i, X_j)$$

$$\text{such that } 1 \le i < j \le n, \quad X_i + X_j > 0\}.$$

$$\xi_1 = E[\psi(X_1, X_2)\psi(X_1, X_3)] - \gamma^2$$

$$= P[X_1 + X_2 > 0, X_1 + X_3 > 0] - \gamma^2$$

and

$$\xi_2 = E[\psi(X_1, X_2)\psi(X_1, X_2)] - \gamma^2$$
$$= \gamma - \gamma^2 = \gamma(1-\gamma).$$

The variance of U-statistic is then

$$Var(U(X_1, \cdots, X_n)) = \frac{1}{\binom{n}{2}}[\binom{2}{1}\binom{n-2}{1}\xi_1 + \binom{2}{2}\binom{n-2}{0}\xi_2]$$
$$= \frac{2}{n(n-1)}[2(n-2)\{P[X_1 + X_2 > 0, X_1 + X_3 > 0]$$
$$-\gamma^2\} + \gamma(1-\gamma)].$$

Asymptotic Variance of U-statistics: Let $U(X_1, \cdots, X_n)$ be the U-statistic for a symmetric kernel $h(X_1, \cdots, X_r)$.

If $E[h^2(X_1, \cdots, X_r)] < \infty$, then

$$\lim_{n \to \infty} nVar[U(X_1, \cdots, X_n)] = r^2 \xi_1. \tag{6.2.7}$$

Proof: Because $E[h^2(X_1, \cdots, X_r)] < \infty$, variance of U-statistic exists. Define

$$K_c = \frac{(r!)^2}{c!((r-c)!)^2}, \quad c = 1, 2, \cdots, r.$$

$$Var(U) = \frac{1}{\binom{n}{r}} \sum_{c=1}^{r} \binom{r}{c}\binom{n-r}{r-c}\xi_c.$$

$$nVar(U) = \frac{n}{\binom{n}{r}} \sum_{c=1}^{r} \binom{r}{c}\binom{n-r}{r-c}\xi_c.$$

The general term in the resulting sum is

$$\frac{n}{n!}(n-r)!r! \times \frac{r!}{(r-c)!c!} \times \frac{(n-r)!}{(r-c)!(n-2r+c)!}\xi_c$$
$$= K_c \frac{n(n-r)(n-r-1)\cdots(n-2r+c+1)}{n(n-1)\cdots(n-r+1)}\xi_c$$

which goes to $r^2\xi_1$ as $n \to \infty$ for $c = 1$ and for $c > 1$, the term goes to zero. Hence the result.

One Sample U-statistic Theorem (Hoeffding, 1948)

Let $X_1, X_2, \cdots, X_n$ be a random sample from a distribution $F \in \mathbb{F}$ and $\gamma = \gamma(F)$ be an estimable parameter of degree r with symmetric kernel $h(X_1, \cdots, X_r)$. Define

$$\text{(i)} \quad U = U(X_1, \cdots, X_n) = \frac{1}{\binom{n}{r}} \sum_{\beta \in B} h(X_{\beta_1}, \cdots, X_{\beta_r}),$$

where B consists of all subsets of r integers chosen without replacement from $\{1, 2, \cdots, n\}$.

(ii) $\xi_1 = \{E[h(X_1, \cdots, X_r)h(X_1, X_{r+1}, \cdots, X_{2r-1})] - \gamma^2\}$.

If $\xi_1 > 0$ then

$$\sqrt{n}(U - \gamma) \xrightarrow{D} N(0, r^2\xi_1).$$

In applications of the above result, ξ_1 is computed using the relation: $\xi_1 = Var[h_1(X_1)]$ where

$$h_1(x) = E[h(x, X_2, \cdots, X_r)].$$

6.3 Tests for Exponentiality

(A) Hollander and Proschan Test (1972)

$H_0 : F(x) = 1 - e^{-\lambda x}; x \geq 0, \lambda > 0$ (unspecified)

$H_1 : F$ is NBU but not exponential.

Let $T_1, T_2, \cdots, T_n$ be a random sample of size n from the distribution F. F is assumed to be continuous.

If F is exponential then it satisfies Cauchy functional equation

$$\overline{F}(s + t) = \overline{F}(s)\overline{F}(t), \quad s, t > 0.$$

If, however, it belongs to the NBU class then

$$\overline{F}(s + t) \leq \overline{F}(s)\overline{F}(t), \quad s, t > 0.$$

Therefore, equivalently one can test

$H_0 : \overline{F}(s + t) = \overline{F}(s)\overline{F}(t)$ against

$H_1 : \overline{F}(s + t) \leq \overline{F}(s)\overline{F}(t)$; with strict inequality for some s, t.

Hollander and Proschan's test (1972) is based on the measure τ defined as

$$\begin{aligned}\tau &= \int_0^\infty \int_0^\infty \{\overline{F}(s)\overline{F}(t) - \overline{F}(s+t)\}dF(s)dF(t) \\ &= \frac{1}{4} - \int_0^\infty \int_0^\infty \overline{F}(s+t)dF(s)dF(t) \\ &= \frac{1}{4} - \gamma \quad \text{(say)}.\end{aligned}$$

The alternative corresponds to the positive values of τ or equivalently, to the small values γ. Now

$$\begin{aligned}\gamma &= \int_0^\infty \int_0^\infty \overline{F}(s+t)dF(s)dF(t) \\ &= \int_0^\infty \int_0^\infty P[T_1 > s+t]dF(s)dF(t) \\ &= P[T_1 > T_2 + T_3]\end{aligned}$$

where T_1, T_2 and T_3 are continuous i.i.d.r.v.s with common distribution F.

Since $T_1, T_2, \cdots, T_n$ is a random sample of size n from the distribution F, the U-statistic estimator of γ based on the kernel

$$h(T_1, T_2, T_3) = \psi(T_1, T_2, T_3) = \begin{cases} 1 \text{ if } & T_1 > T_2 + T_3 \\ 0 & \text{otherwise} \end{cases}$$

is obtained by first constructing the symmetric kernel $(h^*(T_1, T_2, T_3))$:

$$h^*(T_1, T_2, T_3) = \frac{1}{3!}[2\psi(T_1, T_2, T_3) + 2\psi(T_2, T_1, T_3) + 2\psi(T_3, T_1, T_2)].$$

$$\begin{aligned}U &= \frac{6}{n(n-1)(n-2)} \sum_{i<j<k=1}^{n} \sum^{n}\sum^{n} \frac{1}{3}[(\psi(T_i, T_j, T_k)] \\ &\quad +\psi(T_j, T_i, T_k) + \psi(T_k, T_i, T_j)] \\ &= \frac{2}{n(n-1)(n-2)} \sum_{i<j<k=1}^{n} \sum^{n}\sum^{n} \psi(T_{(k)}, T_{(i)}, T_{(j)}).\end{aligned}$$

Now if each T_i is exponential with parameter λ, then $\lambda T_1, \lambda T_2, \cdots, \lambda T_n$ are independent exponentials with parameter 1. Since

$$\begin{aligned}U &= U[T_1, T_2, \cdots, T_n] \\ &= U[\lambda T_1, \lambda T_2, \cdots, \lambda T_n], \quad \lambda > 0;\end{aligned}$$

it follows that U is free of the parameter λ. Under H_0

$$\begin{aligned}\gamma &= P[T_1 > T_2 + T_3] \\ &= \int_0^\infty \int_0^\infty e^{-(s+t)}.e^{-s}e^{-t}dsdt \\ &= \int_0^\infty e^{-2s}ds \int_0^\infty e^{-2t}dt \\ &= \frac{1}{4}.\end{aligned}$$

Hollander and Proschan have obtained the exact null distribution of the statistic

$$S_n = \frac{n(n-1)(n-2)}{2}U_n.$$

The test consists of rejecting H_0 for large values of the statistic S_n or equivalent versions of it. Exact probabilities are computed in special cases and lower and upper percentile points based on Monte Carlo simulations are given for $n = 4(1)20(5)50$. By U-statistic theorem the limiting distribution of $\sqrt(n)(U_n - \gamma)$ is asymptotically normal with mean zero and variance $\frac{5}{432}$. The corresponding test based on standard normal variate is unbiased.

Remark. Since IFR class is contained in NBU class, the test for NBU alternatives focusses on a larger class of alternatives than do the IFR tests.

(B) Tests for exponentiality against positive-ageing based sample spacings

Let $T_1, T_2, \cdots, T_n$ be a random sample of size n from the distribution F which is continuous and have support on the positive half of the real line. Let f be its density, $\overline{F}$, its survival function and r_F, its failure rate function. The total time on test transform (TTT) of F is defined as

$$H_F^{-1}(t) = \int_0^{F^{-1}(t)} \overline{F}(u)du \quad \text{for} \quad 0 \leq t \leq 1;$$

where

$$F^{-1}(t) = \inf\{u : F(u) \geq t\}.$$

Figure 6.1 shows the curve $y = F(x)$ and the shaded area represents $H_F^{-1}(t)$.

In what follows, we shall see how the difference in the nature of TTT transforms of F when (i) F is exponential and (ii) F exhibits positive ageing, can be used to construct tests for exponentiality against positive ageing.

Some useful properties of TTT transform

(i) $H_F^{-1}(t)$ is integral of a non-negative function and hence $H_F^{-1}(t)$ is an $\uparrow$ function of t,

(ii) If $F^{-1}(t) = \infty$ then $H_F^{-1}(t) = \int_0^\infty \overline{F}(u)du = \mu =$ the mean of F. This is the largest value of $H_F^{-1}(t)$.

(iii)

$$
\begin{aligned}
\frac{d}{dt} H_F^{-1}(t) &= \overline{F}(F^{-1}(t)) \frac{d}{dt} F^{-1}(t) \\
&= \overline{F}(F^{-1}(t)) \cdot \frac{1}{[\frac{d}{du} F(u)]_{u=F^{-1}(t)}} \\
&= \overline{F}(F^{-1}(t)) \cdot \frac{1}{f[F^{-1}(t)]} \\
&= \left[\frac{f(F^{-1}(t))}{\overline{F}(F^{-1}(t))}\right]^{-1} = [r(F^{-1}(t))]^{-1},
\end{aligned}
$$

where $r(F^{-1}(t))$ is failure rate evaluated at $F^{-1}(t)$.

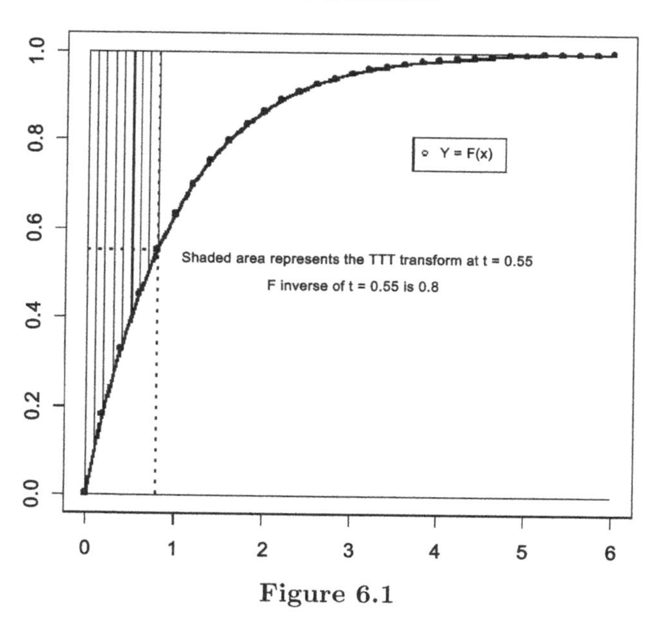

Figure 6.1

(iv) F is IFR if and only if $H_F^{-1}(t)$ is a concave function of t.

$$\begin{aligned} F \quad \text{is} \quad IFR &\Leftrightarrow r(x) \uparrow x \\ &\Leftrightarrow r[F^{-1}(t)] \uparrow t \\ &\Leftrightarrow \{r[F^{-1}(t)\}^{-1} \downarrow t \end{aligned}$$

$$\begin{aligned} F \quad \text{is} \quad IFR &\Leftrightarrow \frac{d}{dt} H_F^{-1}(t) \downarrow t \\ &\Leftrightarrow \frac{d^2}{dt^2}(H_F^{-1}(t)) \leq 0 \\ &\Leftrightarrow H_F^{-1}(t) \ \text{ is a concave function of } t. \end{aligned}$$

(v) F is NBUE iff $\psi_F(t) \geq t, 0 \leq t \leq 1$.

$$\begin{aligned} F \ \text{ is NBUE } \quad &\Leftrightarrow \int_0^\infty \frac{\overline{F}(t+x)}{\overline{F}(t)} dx \leq \mu, \quad \forall \ t \geq 0 \\ &\Leftrightarrow \int_t^\infty \frac{\overline{F}(y)}{\overline{F}(t)} dy \leq \mu \quad \text{by putting } \ x+t=y \\ &\Leftrightarrow \frac{1}{u} \int_{F^{-1}(1-u)}^\infty \overline{F}(y) dy \leq \mu \\ &\quad \text{by writing } \ \overline{F}(t) = u, 0 \leq u \leq 1 \\ &\Leftrightarrow \frac{1}{u}[\mu - \int_0^{\overline{F}^1(1-u)} \overline{F}(y) dy] \leq \mu, \quad 0 \leq u \leq 1 \\ &\Leftrightarrow H_F^{-1}(1-u) \geq \mu(1-u), \quad 0 \leq u \leq 1 \\ &\Leftrightarrow \frac{1}{\mu} H_F^{-1}(1-u) \geq (1-u), \quad 0 \leq u \leq 1 \\ &\Leftrightarrow \psi_F(t) \geq t, \quad 0 \leq t \leq 1. \end{aligned}$$

(vi) Let F be exponential (λ) where $\lambda = \frac{1}{\mu}$

$$\begin{aligned} H_F^{-1}(t) &= \int_0^{F^{-1}(t)} \overline{F}(u) du = \int_0^{F^{-1}(t)} e^{-\lambda u} du \\ &= (-\frac{1}{\lambda} e^{-\lambda u})_0^{F^{-1}(t)} = (\frac{1}{\lambda} - \frac{1}{\lambda} e^{-\lambda(F^{-1}(t))}) \end{aligned}$$

where

$$F^{-1}(t) = -\frac{1}{\lambda} \log(1-t).$$

This gives

$$H_F^{-1}(t) = t/\lambda = \mu t.$$

If F is exponential then $H_F^{-1}(t) = \mu t$, where μ is the mean of the distribution.

Scaled TTT transform of F

$$\psi_F(x) = \frac{H_F^{-1}(x)}{H_F^{-1}(1)} = \frac{1}{\mu}\int_0^{F^{-1}(x)} \overline{F}(u)du.$$

Note that ψ_F is scale invariant. If F is exponential then $\psi_F(x) = x$.

Estimation of TTT transform of F:

$$\begin{aligned}
\hat{F}(t) &= F_n(t) = \text{empirical distribution function} \\
&= \frac{(\#\text{sample } T_i \leq t)}{n} \\
H_{F_n}^{-1}(t) &= \text{Sample TTT transform} \\
&= \hat{H_F^{-1}}(t) \\
&= \int_0^{F_n^{-1}(t)} \overline{F}_n(u)du. \\
H_{F_n}^{-1}(j/n) &= \int_0^{T_{(j)}} \overline{F}_n(u)du \\
&= \sum_{i=1}^{j} \int_{T_{(i-1)}}^{T_{(i)}} [1 - \frac{(i-1)}{n}]du.
\end{aligned}$$

Note that between $T_{(i-1)}$ and $T_{(i)}$, $\hat{F}_n(t)$ and hence $\hat{\overline{F}}_n(t)$ is constant. Therefore, $H_{F_n}^{-1}(t)$ changes at $1/n, 2/n, \cdots$ and is constant in between.

$$\begin{aligned}
H_{F_n}^{-1}(j/n) &= \sum_{i=1}^{j} \frac{(n-i+1)}{n}[T_{(i)} - T_{(i-1)}] \\
H_{F_n}^{-1}(j/n) &= \frac{S_j}{n}, \quad j = 0, 1, 2, \cdots, n
\end{aligned}$$

where

$$S_j = \sum_{i=1}^{j}(n-i+1)(T_{(i)} - T_{(i-1)})$$
$$= \sum_{i=1}^{j} \mathcal{D}_i,$$

where $D_i = (n-i+1)(T_{(i)} - T_{(i-1)})$ are the normalised sample spacings. Therefore

$$H_{F_n}^{-1}(j/n) = \frac{1}{n}\sum_{i=1}^{j}\mathcal{D}_i$$

and estimator of scaled TTT transform

$$\psi_{F_n}(j/n) = \frac{1}{\overline{T}_i}\sum_{i=1}^{j}\left(\frac{n-i+1}{n}\right)(T_{(i)} - T_{(i-1)})$$

where $\overline{T}_n$ is mean of T_i.

$$= \frac{1}{\overline{T}_n}\frac{S_j}{n}$$
$$= \frac{S_j}{S_n}, \quad \text{where } S_n = n\overline{T}_n$$

and $S_0 = 0$.

A simple graphical test of exponentiality against IFR (not exponential) alternative.

Consider the scaled transform

$$\psi_F(t) = \frac{1}{\mu}\int_0^{F^{-1}(t)} \overline{F}(u)du$$

and observe from properties (iv) and (vi) that

(i) F is exponential $\Leftrightarrow \psi_F(t) = t$ and

(ii) F is IFR $\Leftrightarrow \psi_F(t)$ is concave.

Therefore a simple graphical test based on TTT plot is obtained by plotting $U_j = \frac{S_j}{S_n}$ against j/n.

If H_0 is true then the points will fall on or near the line $y = x$ if H_1 is true then the points will fall on or near a concave curve above the line $y = x$. (See Figures 6.2 and 6.3.)

(b) Furthermore observe that by property (v) if F is NBUE then $\psi_F(t) \geq t$.

Remark: It will be difficult to distinguish between the two alternatives (i) IFR and (ii) NBUE on the basis of the TTT plot. However, the TTT plot can be used to get a rough idea about the existence or otherwise of positive ageing phenomenon. Also the above discussion serves the purpose of illustrating how sample spacings can provide a basis for constructing the tests of exponentiality. In general, observe that

$$\psi_{F_n}(j/n) = U_j = \frac{S_j}{S_n} = \frac{\sum_{k=1}^{j}(n-k+1)(T_{(k)} - T_{(k-1)})}{n\overline{T}_n}$$

and define

$$K_n = \sum_{j=1}^{n}(U_j - j/n).$$

Note that $K_n = 0$ if F is expoential and $K_n > 0$ if $F \in$ positive ageing class.

In the sequel we shall discuss an analytical test of exponentiality vs. the alternative that F is NBUE developed by Hollander and Proschan (1975).

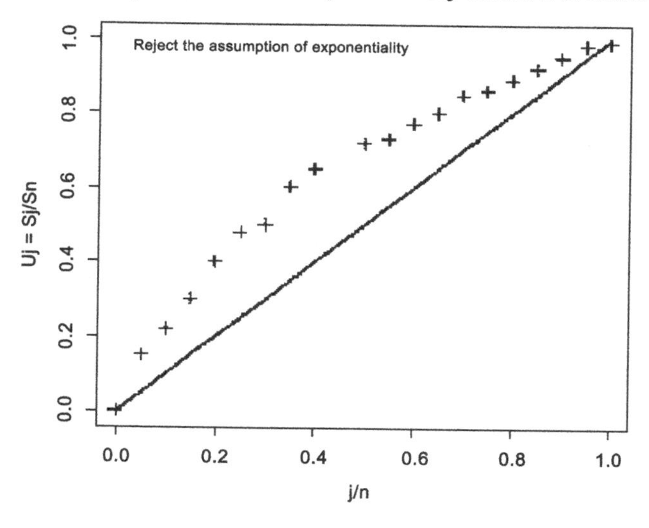

Figure 6.2

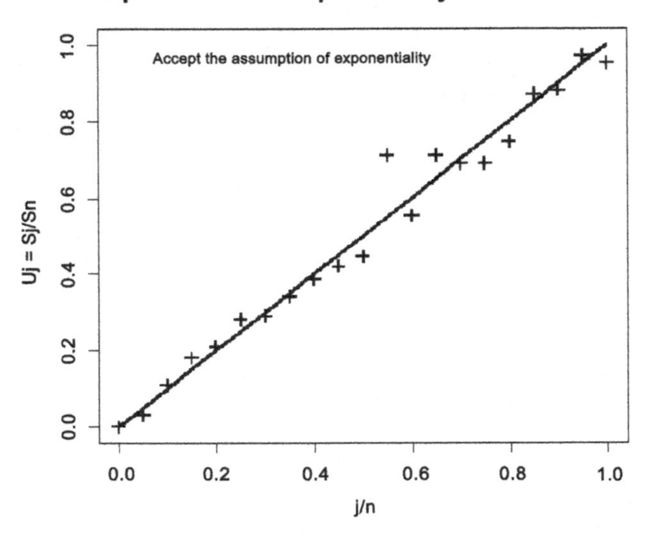

Figure 6.3

(c) Analytical test of exponentiality within NBUE class of alternatives
$H_0 : F(t) = 1 - \exp(-\lambda t), \quad t \geq 0, \lambda > 0$ (unspecified)
H_1: F is NBUE but *not* exponential.
The test is based on $K_n = \sum_{j=1}^{n} (U_j - j/n)$, where

$$\begin{aligned} U_j &= \frac{S_j}{S_n} \\ &= \frac{1}{S_n}\left(\sum_{k=1}^{j} T_{(k)} + (n-j)T_{(j)}\right), \end{aligned}$$

where $\sum_{k=1}^{j} T_{(k)}$ = sum of all observed lifetimes upto j-th failure.

$(n-j)T_{(j)}$ = sum of observed lifetimes of the unfailed items upto j-th failure.

$$K_n = \left[\left\{\frac{1}{S_n}\sum_{j=1}^{n}\sum_{k=1}^{j}(n-k+1)(T_{(k)} - T_{(k-1)})\right\} - \frac{n+1}{2}\right]$$

$$= \{\frac{1}{S_n}\sum_{i=1}^{n}(2n-2i+1)T_{(i)}\} - (\frac{n+1}{2})$$

$$= \frac{1}{\overline{T}_n}[\sum_{i=1}^{n}(3/2 - \frac{2i}{n} + \frac{1}{2n})T_{(i)}]$$

$$(K_n - \frac{1}{2})\overline{T}_n/n = n^{-1}\sum_{i=1}^{n}(3/2 - 2i/n)T_{(i)}.$$

Moore's Theorem (1968)

Let $X_{(1)} \leq X_{(2)} \leq \cdots \leq X_{(n)}$ be the order statistics corresponding to a random sample of size n from a continuous distribution F. Let

$$\tau_n = n^{-1}\sum_{i=1}^{n} L(i/n)X_{(i)}$$

and

$$\sigma^2 = \sigma^2(F) = 2\int_{s<t}\int [L(F(s))][L(F(t))]F(s)[1-F(t)]dsdt \qquad (*)$$

Then

$$n^{1/2}[\tau_n - \int_{-\infty}^{\infty} xL(F(x))dF(x)] \xrightarrow{\mathcal{D}} N(0,\sigma^2)$$

provided

(i) $E|X_1| < \infty$

(ii) $\sigma^2 < \infty$.

(iii) L is continuous function on $[0,1]$ except perhaps at a finite number of jump discontinuities at $a_1, a_2, \cdots, a_M$ and is of bounded variation on $[0,1] - \{a_1, a_2, \cdots a_M\}$.

Let $L(t) = \frac{3}{2} - 2t$. This gives

$$(K_n - 1/2)\frac{\overline{T}_n}{n} = n^{-1}\sum_{i=1}^{n} L(i/n)T_{(i)} = \tau_n.$$

Verification of the conditions

(i) $E[|T_1|] = E[T_1] = \mu = \frac{1}{\lambda} < \infty$ under H_0

(ii) $\sigma^2(F) = 2\iint_{s<t} L[F(s)]\ L(F(t))F(s)\overline{F}(t)dsdt.$

This simplifies to

$$\sigma^2(F) = \frac{1}{12\lambda^2} < \infty.$$

(iii) $L(t) = 3/2 - 2t$ is continuous function on $[0, 1]$.
Therefore by Moore's Theorem, under H_0:

$$\sqrt{n}[(K_n - 1/2)\frac{\overline{T}_n}{n} - \int_0^\infty x(3/2 - 2F(x))dF(x)] \xrightarrow{\mathcal{D}} N(0, \sigma^2). \quad (6.3.1)$$

But under H_0:

$$\begin{aligned}\mu(L, F) &= \int_0^\infty x(3/2 - 2F(x))dF(x) \\ &= \int_0^\infty x(3/2 - 2(1 - e^{-\lambda x}))d(1 - e^{-\lambda x}) \\ &= \lambda[-\frac{1}{2\lambda^2} + \frac{1}{2\lambda^2}] = 0.\end{aligned}$$

Therefore under H_0:

$$\sqrt{n}[\frac{\overline{T}_n}{n}(K_n - 1/2)] \xrightarrow{\mathcal{D}} N(0, \sigma^2)$$

or

$$\frac{\sqrt{n}(\frac{1}{n}(K_n - 1/2))}{\sigma/\overline{T}_n} \xrightarrow{\mathcal{D}} N(0, 1) \quad \text{under} \quad H_0. \quad (6.3.2)$$

But $\overline{T}_n$ is the consistent estimator of μ, the mean of F. That is, $\frac{\overline{T}_n}{\mu} \xrightarrow{P} 1$.
Therefore by Slutsky's theorem,

$$\frac{\sqrt{n}(\frac{1}{n}(K_n - 1/2))}{(\sigma/\mu)} \xrightarrow{\mathcal{D}} N(0, 1) \quad \text{under} \quad H_0. \quad (6.3.3)$$

Note that

$$\sigma^2 = \frac{\mu^2}{12} \Rightarrow \frac{\sigma}{\mu} = \sqrt{\frac{1}{12}}.$$

Hence the test is: Reject H_0 if the computed value of

$$\frac{n^{-1/2}(K_n - 1/2)}{\sqrt{\frac{1}{12}}} = Z_0 \geq z_{(1-\alpha)}$$

where $z_{(1-\alpha)}$ is such that $P[SNV \leq z_{(1-\alpha)}] = 1 - \alpha$.

For small sample sizes, the cutoff points for total time on test statistic are available (Barlow, 1968).

Consistency of the Test: A test is consistent if the power of the test tends to 1 as $n \to \infty$. That is, for the problem under consideration,

$$\lim_{n\to\infty} P_1 \left[\left(\frac{n^{-1/2}(K_n - 1/2)}{\sqrt{\frac{1}{12}}} \right) \geq z_{(1-\alpha)} \right] = 1,$$

where P_1 denotes probability under H_1.

Theorem 6.3.1. Suppose under H_0, $\sqrt{n}\frac{(T_n-\mu_n)}{(\sigma_{0n})} \xrightarrow{\mathcal{D}} N(0,1)$ where μ_n is the mean of T_n and σ_{0n} is its standard deviation under H_0. If, under H_1,

(i) $E_{H_1}(T_n) = \theta_n > \mu_n$ and $\theta_n - \mu_n \xrightarrow{P} c > 0$,

(ii) $Var_{H_1}(T_n) = \sigma_{1n}^2 \not\to \infty$ as $n \to \infty$ and $\frac{\sigma_{1n}^2}{\sigma_{0n}^2} \to a \neq \infty$ and

(iii) $\sqrt{n}\frac{(T_n-\theta_n)}{\sigma_{1n}} \xrightarrow{\mathcal{D}} N(0,1)$ under H_1,

then the test is consistent.

Proof

$$P_{H_1}\left[\frac{\sqrt{n}(T_n-\mu_n)}{\sigma_{0n}} > z_{(1-\alpha)}\right]$$
$$= P_{H_1}\left[\frac{\sqrt{n}(T_n-\theta_n)}{\sigma_{1n}} > z_{(1-\alpha)}\frac{\sigma_{0n}}{\sigma_{1n}} - \frac{\sqrt{n}(\theta_n-\mu_n)}{\sigma_{1n}}\right].$$

As $n \to \infty, \theta_n - \mu_n \xrightarrow{p} c > 0$ and $\sigma_{1n}^2 \not\to \infty$.

Therefore, $\frac{\sqrt{n}(\theta_n-\mu_n)}{\sigma_{1n}} \to -\infty$. Moreover, $\frac{\sigma_{1n}^2}{\sigma_{0n}^2} \to a \neq \infty$. Therefore

$$P_{H_1}\left[\frac{\sqrt{n}(\theta_n-\mu_n)}{\sigma_{1n}} > z_{(1-\alpha)}\frac{\sigma_{0n}}{\sigma_{1n}} - \frac{\sqrt{n}(\theta_n-\mu_n)}{\sigma_{1n}}\right]$$

$$\to P[SNV > -\infty] = 1 \quad \text{as} \quad n \to \infty.$$

Hence the test is consistent.

We apply the above theorem to the problem under consideration. It is enough to show that

$$\mu(L,F) = \int_0^\infty xL[\{F(x)\}]dF(x) \geq 0$$

with equality under H_0 and strict inequality under H_1.

First consider the class of HNBUE distributions

$$\begin{aligned}\mu(L,F) &= \int_0^\infty x(3/2 - 2F(x))dF(x) \\ &= 3/2\mu + 2\int_0^\infty x(1-F(x))dF(x) - 2\int_0^\infty xdF(x) \\ &= -\frac{\mu}{2} + 2\int_0^\infty x\overline{F}(x)dF(x).\end{aligned}$$

$$\begin{aligned}\mu(L,F) &= -\mu/2 - (-2\frac{x}{2}\overline{F}^2(x))_0^\infty + \int_0^\infty [\overline{F}(x)]^2 dx \\ &= -\mu/2 + \int_0^\infty (\overline{F}(x))^2 dx. \qquad (6.3.4)\end{aligned}$$

To show that for F HNBUE,

$$\int_0^\infty (\overline{F}(x))^2 dx \geq \mu/2. \qquad (6.3.5)$$

Proof.

$$F \text{ HNBUE} \quad \Leftrightarrow \int_t^\infty \overline{F}(x)dx \leq \int_t^\infty \overline{G}(x)dx;$$

where $\overline{G}(x) = e^{-\lambda x}, \quad x > 0.$

$$\Leftrightarrow \int_0^t \overline{F}(x)dx \geq \int_0^t \overline{G}(x)dx, \qquad (6.3.6)$$

provided F and G have the same means. To this end, we shall use the following theorem.

Theorem [Barlow and Proschan (1975), p. 121]

Let

$$\int_0^t \overline{F}_i(x)dx \geq \int_0^t \overline{G}_i(x)dx \quad \text{for all} \quad t > 0$$

and F and G have the same means, then

$$\int_0^t \pi_{i=1}^n \overline{F}_i(x)dx \geq \int_0^t \pi_{i=1}^n \overline{G}_i(x)dx \quad \text{for all} \quad t > 0. \qquad (6.3.7)$$

From the theorem,

$$\int_0^t (\overline{F}(x))^2 dx \geq \int_0^t (\overline{G}(x))^2 dx.$$

Now take limits as $t \to \infty$,

$$\lim_{t\to\infty}\int_0^t (\overline{F}(x))^2 dx \geq \lim_{t\to\infty}\int_0^t (\overline{G}(x))^2 dx.$$

That is

$$\int_0^\infty (\overline{F}(x))^2 dx \geq \int_0^\infty e^{-2\lambda x} dx = \frac{1}{2\lambda} = \mu/2.$$

Thus, the test is consistent for HNBUE but not exponential class of alternatives and hence for the NBUE but not exponential class of alternatives as well.

(d) The Klefsjo Test of Exponentiality against IFRA alternative based on scaled TTT transform.

The class of IFRA distributions plays a central role in the statistical theory of reliability. It is the smallest class which contains the exponential distribution and is closed under formation of coherent systems of independent components. Hence testing for exponentiality within this class is of practical importance.

$$H_0 : F(x) = 1 - \exp(-\lambda x),$$
$$x \geq 0, \lambda > 0 \quad \text{(unspecified)}$$

$$H_1 : F \text{ is IFRA but not exponential.}$$

This test is based on a necessary (but not sufficient) condition for F to be IFRA. The statement of the condition is:

If F is a life distribution which is IFRA then $\frac{\psi_F(t)}{t}$ is decreasing function of t.

Consequently if we have a random sample from IFRA distribution then we expect $\frac{U_j}{j/n}$ to decrease as j increases. Note that U_j is as defined in Hollander and Proschan test (1975). This means that

$$\frac{U_i}{i/n} > \frac{U_j}{j/n} \quad \text{for } j > i \text{ and } i = 1, 2, \cdots, n-1.$$

Multiplication by ij/n and summation over i and j gives the test statistic

$$B = \sum_{i=1}^{n-1}\sum_{j=i+1}^{n} (ju_i - iu_j). \tag{6.3.8}$$

We expect positive values of B if F is IFRA. Expression (6.3.8) on simplification yields,

$$B = \sum_{j=1}^{n}(\beta_j D_j / S_n)$$

where

$$\beta_j = \frac{1}{6}[2j^3 - 3j^2 + j(1 - 3n + 3n^2) + 2n + 3n^2 + n^3]. \tag{6.3.9}$$

The exact distribution of B

Let $T_{(1)} < T_{(2)} \cdots < T_{(n)}$ be an ordered sample from exponential distribution. Because of the scale invariance, we can assume that $\lambda = 1/2$ (see Barlow and Proschan (1975), p. 59). By using a technique similar to the one given in Langenberg and Srinivasan (1979) we get

$$P[B > t] = \sum_{j=1}^{n} \sum_{\substack{i=1 \\ i \neq j}}^{n} \{(\beta_j - t)/(\beta_j - \beta_i)\}\delta_j$$

where

$$\delta_j = \begin{cases} 1 \text{ if } \ \beta_j > t \\ 0 \text{ otherwise.} \end{cases}$$

It may be noted that if there is a tie between β_j's a more complicated expression results after using Theorem 2.4 of Box (1954).

Critical (exact) values of the statistic $B\sqrt{\frac{210}{n^5}}$ are provided by Klefsjö (1983) for sample sizes 5(5) 75.

The Asymptotic Distribution

It can be shown that B is asymptotically normally distributed under general assumptions.

Define

$$\begin{aligned} \mu(F) &= \int_0^\infty \overline{F}(x)dx \\ \mu(J_B, F) &= \int_0^\infty xJ_B(F(x))dF(x) \\ \sigma^2(J_B, F) &= \int_0^\infty \int_0^\infty J_B(F(x))J_B(F(y)) \\ &\quad \times \{F(\min(x, y) - F(x)F(y)) \times dxdy\}. \end{aligned}$$

It can be proved that

$$\left\{\sqrt{n}\left(\frac{B}{n^3} - \frac{\mu(J_B, F)}{\sigma(J_B, F)}\right)\right\} \xrightarrow{\mathcal{D}} N(0,1),$$

where

$$J_B(u) = \frac{1}{3}(2 - 3u - 3u^2 + 4u^3). \tag{6.3.10}$$

We get

$$(B\sqrt{\frac{210}{n^5}}) \xrightarrow{D} N(0,1). \tag{6.3.11}$$

The test based on (6.3.11) is consistent against the class of continuous IFRA life distributions since $\mu(J_B, F_0) = 0$ and $\mu(J_B, F) > 0$ where F_0 is exponential and F is IFRA (not exponential) continuous distribution.
(e) Deshpande's Test: A class of tests for exponentiality against increasing failure rate average alternatives (Deshpande (1983)).

$$H_0 : F(x) = 1 - \exp(-\lambda x), \quad x \geq 0, \lambda > 0 \;\; \text{(unspecified)}$$

$$H_1 : F \;\text{ is IFRA but not exponential.}$$

Rationale of the test: It has been shown (p. 18) that

$$F \text{ is IFRA} \quad \Leftrightarrow [\overline{F}(x)]^b \leq [\overline{F}(bx)], \quad 0 \leq b \leq 1, \quad 0 \leq x < \infty. \tag{6.3.12}$$

Equality in (6.3.12) holds iff F is exponential. For F, not exponential, but in IFRA class,

$$[\overline{F}(x)]^b < \overline{F}(bx), \quad 0 < b < 1, \quad 0 \leq x < \infty. \tag{6.3.13}$$

Let

$$M_F = \int_0^\infty \overline{F}(bx)dF(x).$$

Under H_0,

$$\gamma = M_F = \frac{1}{(b+1)} \;\; \text{for} \;\; 0 < b < 1.$$

Under H_1,

$$\gamma > \frac{1}{(b+1)}.$$

For a chosen number b between 0 and 1 (0.5 and 0.9 are possible choices), $(\gamma - \frac{1}{b+1})$ may be taken as a measure of divergence of F from exponentiality. *Construction of U-statistic for the testing problem*: Let $X_1, \cdots, X_n$ be a random sample from the distribution F.

Let $\tau = \gamma - \frac{1}{(b+1)}$ and

$$h(X_1, X_2) = \psi(X_1 - bX_2) = \begin{cases} 1 \text{ if } X_1 > bX_2 \\ 0 \text{ otherwise} \end{cases}$$

$$\begin{aligned} E_{H_0}(\psi(X_1 - bX_2)) &= P[X_1 > bX_2] \\ &= \int_0^\infty P[X_1 > bx]dF(x) \\ &= \int_0^\infty \overline{F}(bx)dF(x) \\ &= \int_0^\infty (\overline{F}(x))^b dF(x) \\ &= \frac{1}{(b+1)} \\ &= \gamma. \end{aligned}$$

Thus γ is an estimable function of degree 2 and $h(X_1, X_2)$ is a kernel of degree 2. However, $h(X_1, X_2)$ is not symmetric. Hence a symmetric kernel is obtained as follows:

$$h^*(X_1, X_2) = \frac{1}{2}[\psi(X_1 - bX_2) + \psi(X_2 - bX_1)].$$

Using this symmetric kernel, the corresponding U-statistic is constructed to test the hypothesis of interest.

$$\begin{aligned} U = J_b &= \frac{1}{\binom{n}{2}} \cdot \frac{1}{2} \sum_{i<j}\sum [\psi(X_i - bX_j) + \psi(X_j - bX_i)] \\ &= \frac{1}{n(n-1)} \sum_{i<j}\sum [\psi(X_i - bX_j) + \psi(X_j - bX_i)]. \end{aligned}$$

$E(U) = \gamma$ under H_0 and asymptotic variance of $\sqrt{n}(U - \gamma)$ is $4\xi_1$. Under H_0, ξ_1 is given by

$$\xi_1 = \frac{1}{4}\left\{1 + \frac{b}{2+b} + \frac{1}{2b+1} + \frac{2(1-b)}{(1+b)} - \frac{2b}{(1+b+b^2)} - \frac{4}{(b+1)^2}\right\}.$$

Asymptotic Distribution: By the one-sample U-statistic theorem,

$$Z = \frac{\sqrt{n}(J_b - \frac{1}{b+1})}{2\sqrt{\xi_1}} \xrightarrow{a} N(0,1).$$

Small Samples: In this case, H_0 is rejected if $J_b \geq c_{\alpha,n}$ where $c_{\alpha,n}$ is exact critical point such that the test has required size α.

It may be noted that the value of statistic ranges from $\frac{1}{2}$ to 1. It is equal to $\frac{1}{2}$ if $X_{(i)} < bX_{(i+1)}$ for $i = 1, 2, \cdots, n-1$, and is one if $X_{(1)} > bX_{(n)}$. Results of Monte Carlo study are provided in Biometrika (1983). Critical points for $n(n-1)J_b$ for the two commonly used levels, 1% and 5% for $b = 0.5$ and $b = 0.9$ are tabulated. The sample size ranges from 5(2) 15.
Computation of test statistic: The statistic J_b is easy to compute. Multiply each observation by chosen value of b. Arrange $X_1, X_2, \cdots, X_n$ and $bX_1, bX_2, \cdots, bX_n$ together in increasing order of magnitude. Let R_i be the rank of X_i among combined ranking and $J(i)$ be its rank among $X_1, \cdots, X_n$. Then $R_i - j(i) - 1$ is the number of pairs in which X_i is bigger than bX_j. Let

$$\begin{aligned} S &= \sum_i R_i - \sum_i j(i) - n \\ &= \sum_i R_i - \frac{n(n+3)}{2}. \end{aligned}$$

Then

$$J_b = \{n(n-1)\}^{-1} S.$$

It may be noted that this is the Wilcoxon statistic for the data of X's and bX's.
Consistency of the test: The test is consistent whenever $E(J_b) > \frac{1}{(b+1)}$ and the alternatives are continuous increasing failure rate average distributions. In fact if we choose $b = \frac{1}{k}$, where k is an integer at least 2 then this test is consistent against the larger class "new better than used" distributions also.
Asymptotic Relative Efficiency (ARE)

Suppose for the problem of testing of a simple hypothesis that the value of the parameter θ is θ_0 against one of the alternatives that $\theta > \theta_0, \theta < \theta_0$ and $\theta \neq \theta_0$, two or more tests are available. In such a case, Asymptotic Relative Efficiency (ARE) is used to choose one test from the several available tests.

The Pitman-Noether ARE: Let $\beta_N^{(1)}(\theta)$ and $\beta_N^{(2)}(\theta)$ be the power functions of the two tests based on the same set of observations. Assume that both tests are of the same level α. Consider a sequence of alternatives θ_N and sequence $N^* = h(N)$ such that

$$\lim_{N\to\infty} \beta_N^{(1)}(\theta_N) = \lim_{N\to\infty} \beta_{N^*}^{(2)}(\theta_N)$$

where the two limits exist and are neither zero nor one. Then the ARE of test $A^{(2)}$ with respect to test $A^{(1)}$ is defined as

$$e(A^{(2)}, A^{(1)}) = \lim_{N\to\infty} \frac{N}{N^*},$$

provided the limit exists and is independent of the particular sequences $\{\theta_N\}$ and $\{h(N)\}$. For further discussion of ARE refer to Puri and Sen (1971, pp. 112-124).

Asymptotic Relative Efficiency of J_b tests: For two members of the J_b class of tests, namely $b = 0.5$ and $b = 0.9$ ARE is computed for three parametric families of distributions.

The values of ARE of these two tests ($b = 0.5, b = 0.9$) with respect to Hollander and Proschan's (1972) test for the three parametric families Weibull Makeham and linear failure rate are given in Table 6.1. These parametric families depend upon unknown parameter θ in such a way that $\theta = \theta_0$ yields a distribution belonging to the null hypothesis whereas $\theta > \theta_0$ yields distributions from the alternative.

(i) the Weibull distribution,

$$\overline{F}_\theta(x) = \exp(-x^\theta), \quad (\theta \geq 1, x > 0, \theta_0 = 1);$$

(ii) the Makeham distribution,

$$\overline{F}_\theta(x) = \exp\{-[x + \theta(x + e^{-x} - 1)]\}$$
$$(\theta \geq 0, x > 0, \theta_0 = 0);$$

(iii) the linear failure rate distribution,

$$\overline{F}_\theta(x) = \exp\{-(x + \frac{1}{2}\theta x^2)\}, \quad (\theta \geq 0, x > 0, \theta_0 = 0).$$

The table indicates that these tests have high efficiency when compared with the competitor.

Table 6.1

Asymptotic relative efficiency of J_b tests for $b = 0.5, 0.9$.

Distribution	Weibull	Makeham	Linear failure rate
$J_{0.5}$	1.005	0.945	0.931
$J_{0.9}$	1.022	1.020	1.020

The literature abounds with tests of exponentiality. References to some more tests of exponentiality may be found in Bickel and Doksum (1969). Koul (1978) suggested tests based on Koul-Ogorov type distances. These and the tests by Barlow and Campo (1975) are included in Hollander and Proschan's survey (1984) of tests of exponentiality. A class of tests of exponentiality against various alternatives viz. IFRA, NBU and HNBUE is discussed in Deshpande and Kochar (1985). The test statistic is a linear function of order statistic of a random sample. This is the sample version of a functional which discriminates between exponentiality and the alternatives stated above. Kochar (1985) suggested a test for IFRA alternative which is based on the difference statistic $(s \log \overline{F}(t) - t \log \overline{F}(s)), 0 \leq s \leq t < \infty$ which is zero under H_0 and positive under H_1. A test for exponentiality against HNBUE class of alternatives is by Singh and Kochar (1986) which is the extension of Doksum and Yandell (1984) test.

Illustration 6.3: (Deshpande, Gore and Shanubhogue, 1995).

The following table illustrates the computational procedure of $J_{0.5}$ statistic.

Table 6.2

Computation of $J_{0.5}$ statistic

i	ordered X_i	Ordered $\frac{1}{2}X_i$	R_i
1	300	150	2
2	650	325	6
3	800	400	7
4	1280	640	12
5	1710	855	17
6	1920	960	19
7	2050	1025	20
8	2200	1100	22
9	2600	1300	23
10	2950	1475	25

Table 6.2 Cont'd

i	ordered X_i	Ordered $\frac{1}{2}X_i$	R_i
11	3150	1575	26
12	3400	1700	27
13	3500	1750	28
14	4350	2175	30
15	5700	2850	31
16	8100	4050	32
			327

$J_{1/2} = \frac{1}{16 \times 15}[327 - 152] = 0.7292.$
Value of asymptotic normal statistics is: 1.891969.

We see that in this case, the $J_{1/2}$ test rejects the null - hypothesis of exponentiality at the 5% level of significance.
Illustration 6.4: Carry out

(A) Simple graphical test of exponentiality against IFR (not exponential alternatives),
(B) Analytical test of exponentiality within NBUE class of alternatives, for the data of aluminium coupon of exercise 4.2.

Solution: (A) Figure 6.4 shows the plot of s_j/s_n against j/n for the data under consideration.

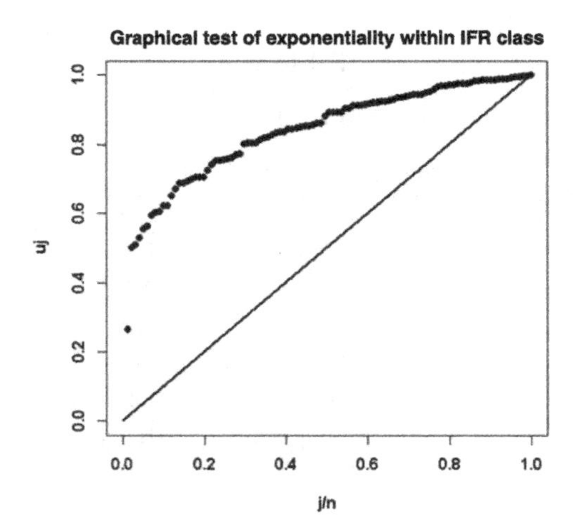

Figure 6.4

We clearly see that the points are on concave curve above the line $y = x$. Hence the exponentiality is rejected and IFR alternative is accepted.

The value of asymptotic standard normal test statistics for analytical test of exponentiality within NBUE class of alternatives is 11.5833 which is larger than 1.64.

The distribution is NBUE (not exponential).

For R-commands see the appendix of this chapter.

Exercises

Exercise 6.1: For the data sets:

(1) data on earthquakes of illustration 4.1, and
(2) data of ball bearing failures of illustration 5.1, carry out:

  (A) Simple graphical test of exponentiality against IFR (not exponential alternatives),
  (B) Analytical test of exponentiality within NBUE class of alternatives,
  (C) Deshpande's test of exponentiality against increasing failure rate average alternatives with $b = 0.5$.

Write your conclusions.

Exercise 6.2: Show that Deshpande's test of exponentiality with increasing failure rate average alternatives is consistent against the larger class of "new better than used" distributions if we choose $b = \frac{1}{k}$ where k is an integer at least 2.

Exercise 6.3: For the data of illustration 6.4, apply Deshpande's test of exponentiality against increasing failure rate average alternatives with $b = 0.5$.

Write your conclusions.

Exercise 6.4: For the data of Illustration 5.1 apply:

(A) Simple graphical test of exponentiality against IFR (not exponential alternatives),
(B) Analytical test of exponentiality within NBUE class of alternatives,
(C) Deshpande's test of exponentiality against increasing failure rate average alternatives with $b = 0.5$.

Write your conclusions.

Appendix

Computation of Jb statistics:

```
> x<-c(300, 650, 800, 1280, 1710, 1920, 2050, 2200, 2600,
 2950, 3150, 3400, 3500, 4350, 5700, 8100);

> y<-x/2;
> z<-c(x,y);
> cr<-rank(z);
> cr;
 [1]  2  6  7 12 17 19 20 22 23 25 26 27 28 30 31 32  1  3
 [19] 4  5  8  9 10 11 13 14 15 16 18 21 24 29
> n<-length(z);
> n;
[1] 32
> r1<-cr[1:(n/2)];
> r1;
 [1]  2  6  7 12 17 19 20 22 23 25 26 27 28 30 31 32
> jb<-data.frame(x,y,r1);

> jb;
       x     y  r1
1    300   150   2
2    650   325   6
3    800   400   7
4   1280   640  12
5   1710   855  17
6   1920   960  19
7   2050  1025  20
8   2200  1100  22
9   2600  1300  23
10  2950  1475  25
11  3150  1575  26
12  3400  1700  27
13  3500  1750  28
14  4350  2175  30
15  5700  2850  31
16  8100  4050  32
```

```
> m<-length(x);
> estjb<-(1/(m*(m-1)))*(sum(r1)-m*(m+3)/2);
> estjb;
[1] 0.7291667
> z1<-1/4*((1+0.5/2.5)+1/2+(2*.5/1.5)-1/(1+.5+.5^2)-4/(.5+1)^2);
> zo<-(sqrt(m))*(estjb-1/(.5+1))/(2*sqrt(z1));
> zo;
[1] 1.891969
```

R-commands for Illustration 6.4:

Graphical Test for exponentiality within NBUE class.

```
> X1<-c(370, 706, 716, 746, 785, 797, 844, 855, 858, 886,
  886,  930, 960, 988, 990, 1000, 1010, 1016, 1018, 1020);

> X2<-c(1055,1085,1102,1102,1108,1115,1120,1134,1140,1199,
  1200, 1200,1203,1222,1235,1238,1252,1288,1262,1269);

> X3<-c(1270,1290,1293,1300,1310,1313,1315,1330,1335,1390,
  1416, 1419,1420,1420,1450,1452,1475,1478,1481,1485);

> X4<-c(1502,1505,1513,1522,1522,1530,1540,1560,1567,1578,
  1594, 1602,1604,1608,1630,1642,1674,1730,1750,1750);

> X5<-c(1763,1768,1781,1782,1792,1820,1868,1881,1890,1893,
  1895, 1910,1923,1940,1945,2023,2100,2130,2215,2268,2440);

> t<-c(X1,X2,X3,X4,X5);
> t<-sort(t);
> t;     # output is not shown;
> n<-length(t);
> D<-numeric(n);
> D[1]<-n*t[1];
> for(i in 2:n)
 {
 D[i]<-(n-i+1)*(t[i]-t[i-1])
 }
> print(D);     # output is not shown;
```

```
> cD<-cumsum(D);
> print(cD);    # output is not shown;
> tbar<-mean(t);
> tbar;
[1] 1401.01
> sn<-n*tbar;
> u<-numeric(n);
> for (j in 1:n)
 {
 u[j]<-cD[j] / sn
 }
> print(u);     # output is not shown;
> sn;
[1] 141502
> jn<-numeric(n);
> for(j in 1:n)
 jn[j]<-j/n
> print(jn);     # output is not shown
> par(font=2,font.axis=2,font.lab=2);

> plot(jn, u, pch=16, xlim = range(0,1), ylim = range(0,1),
   xlab="j/n",ylab="uj", main= "Graphical test");

> x<-c(0,1);
> y<-c(0,1);
> lines(x,y,lwd=2);

> kn<-sum(u-jn);
> zo<-sqrt(12/n)*(kn-.5); zo;
[1] 11.58353     # which is > 1.64;
```

Chapter 7

Two-sample Non-parametric Problem

7.1 Introduction

To begin with we provide an introduction to the U-statistic technique to obtain estimators of two-sample parameters and tests. The technique described for one-sample problems extends directly to the two-sample problem. A parameter $\gamma(F_1, F_2)$ is said to be estimable of degree (r, s) for distributions (F_1, F_2) in a family $\mathbb{F}$ if r and s are the smallest sample sizes from F_1 and F_2 respectively for which there exists an unbiased estimator of γ, based on these observations. In other words, there exists a kernel function $h(\cdot)$ such that

$$E_{(F_1,F_2)}[h(X_1, \cdots, X_r; Y_1, \cdots, Y_s)] = \gamma \quad \text{for every} \quad (F_1, F_2) \in \mathbb{F}.$$

Without any loss of generality we can assume that the two-sample kernel function $[h(\cdot)]$ is symmetric in the observations from each of the two samples. Then the two-sample U-statistic has the form,

$$\begin{aligned} &U(X_1, \cdots, X_m; Y_1, \cdots, Y_n) \\ &= \frac{1}{\binom{m}{r}\binom{n}{s}} \sum_{\alpha \in A} \sum_{\beta \in B} h(X_{\alpha_1}, \cdots, X_{\alpha_r}; Y_{\beta_1}, \cdots, Y_{\beta_s}) \end{aligned}$$

where $(X_1, X_2, \cdots, X_m)$ and $(Y_1, \cdots, Y_n)$ are independent random samples from the distributions F_1 and F_2 respectively and $A(B)$ is the collection of all subsets of $r(s)$ integers chosen without replacement from integers $\{1, \cdots, m\}(\{1, 2, \cdots, n\})$. It is obvious that $U(.,.)$ is an unbiased estimator of γ. The following theorem provides its asymptotic distribution.

Two-sample U-statistic Theorem

Let $X_1, \cdots, X_m$ and $Y_1, \cdots, Y_n$ denote independent random samples from populations with c.d.f.s $F(x)$ and $G(y)$ respectively. Let $h(\cdot)$ be a symmetric kernel for an estimable parameter, γ, of degree (r, s). If $E[h^2(X_1, \cdots, X_r; Y_1, \cdots, Y_s)] < \infty$ and $N = m + n$ then

$$\sqrt{N}[U(X_1, \cdots, X_m; Y_1, \cdots, Y_n) - \gamma]$$

has a limiting normal distribution with mean zero and variance

$$\left[\frac{r^2\xi_{10}}{\lambda} + \frac{s^2\xi_{01}}{1-\lambda}\right], \quad \text{where } 0 < \lambda = \lim_{N\to\infty} \frac{m}{N} < 1.$$

In the above expression,

$$\begin{aligned}\xi_{c,d} = Cov[&h(X_1, \cdots, X_c, X_{c+1}, \cdots, X_r; Y_1, \cdots, Y_d, Y_{d+1}, \cdots Y_s),\\ &h(X_1, \cdots, X_c, X_{r+1}, \cdots, X_{2r-c}; Y_1, \cdots, Y_d, Y_{s+1}, \cdots, Y_{2s-d})].\end{aligned}$$

7.2 Complete or Uncensored Data

Let $X_1, X_2, \cdots, X_m$ and $Y_1, Y_2, \cdots, Y_n$ be two random samples from the distributions F_1 and F_2 respectively. Assume that F_1 and F_2 are continuous and the samples are independent. We wish to test the hypothesis

$$H_0 : F_1(x) = F_2(x) \quad \forall \ x$$

against one of the alternatives:

(i) $H_{11} : \overline{F}_1(x) \geq \overline{F}_2(x)$ with strict inequality for at least one x.

(ii) $H_{12} : \overline{F}_1(x) \leq \overline{F}_2(x)$ with strict inequality for at least one x.

(iii) $H_{13} : \overline{F}_1(x) \neq \overline{F}_2(x)$.

The tests for the above problem suggested independently by Wilcoxon (1945) and Mann-Whitney (1947) in their pioneering papers, turn out to be equivalent.

Define

$$U_{ij} = \begin{cases} 1 & \text{if } X_i > Y_j \\ 0 & \text{if } X_i = Y_j \\ -1 & \text{if } X_i < Y_j \end{cases}$$

and let

$$U = \sum_{i=1}^{m} \sum_{j=1}^{n} U_{ij}.$$

$$E_{H_0}(U_{ij}) = P[X_i > Y_j] - P[X_i < Y_j] = \int_0^\infty \overline{F}(y)dF(y) - \int_0^\infty F(y)dF(y) = 0. \quad (7.2.1)$$

Variance of U under H_0

$$Var_{H_0}(U) = E_{H_0}(\sum_i \sum_j U_{ij})^2, \quad \text{since} \quad E_{H_0}(U) = 0$$

$$= E_{H_0}[\sum_i \sum_j U_{ij}^2 + \sum\sum_{(i,j)\neq(k,\ell)} U_{ij}U_{k\ell}]. \quad (7.2.2)$$

$$\begin{aligned} E_{H_0}(U_{ij}^2) &= P[X_i > Y_j] + P[X_i < Y_j] \\ &= \int_0^\infty \overline{F}(y)dF(y) + \int_0^\infty F(y)dF(y) \\ &= 1. \end{aligned} \quad (7.2.3)$$

$$E_{H_0}[U_{ij}, U_{k\ell}] = \frac{1}{3}, \text{ for } (i = k, j \neq \ell), \text{ [and for } i \neq k, j = \ell \text{ also]}. \quad (7.2.4)$$

Further calculations give

$$\begin{aligned} Var_{H_0}(U) &= E_{H_0}[\sum_i \sum_j U_{ij}^2 + \sum\sum_{(i,j)\neq(k,\ell)} U_{ij}U_{k\ell}] \\ &= \frac{mn}{3}(m + n + 1). \end{aligned} \quad (7.2.5)$$

Therefore under H_0:

$$Z = \frac{U}{\sqrt{\frac{mn(m+n+1)}{3}}} \xrightarrow{\mathcal{D}} N(0, 1) \text{ as } m, n \to \infty.$$

The appropriate tests for the three alternatives are
(i) Reject H_0 in favour of $H_{11} : \overline{F}_1(x) \geq \overline{F}_2(x)$ if $Z_0 \geq z_{(1-\alpha)}$.
(ii) Reject H_0 in favour of $H_{12} : \overline{F}_1(x) \leq \overline{F}_2(x)$ if $Z_0 \leq z_{\alpha}$.
(iii) Reject H_0 in favour of $H_{13} : \overline{F}_1(x) \neq \overline{F}_2(x)$ if $|Z_0| \geq z_{(1-\alpha/2)}$.

Computation of test statistic

$$U = 2R_1 - m(m + n + 1)$$

where R_1 is the rank sum of the first sample in the combined increasing order. This gives,

$$R_1 = \frac{m(m+n+1)}{2} + \frac{U}{2}.$$

Thus U (Mann-Whitney) and R_1 (Wilcoxon) statistics are linearly related.

Tables of exact null distribution of R_1 based on rank order probabilities are available for small (and moderate) sample sizes.

7.3 Randomly Censored (right) Data

Let $X_1, X_2, \cdots, X_m$ be i.i.d. each with distribution F_1 and $C_1, C_2, \cdots, C_m$ be i.i.d. each with distribution G_1 where C_i is the censoring variable associated with X_i. X's and C's are independent. We observe (T_1, ϵ_1), $(T_2, \epsilon_2), \cdots, (T_m, \epsilon_m)$ where $T_i = \min(X_i \leq C_i)$ and $\epsilon_i = I(X_i \leq C_i)$. For the second sample which is independent of $X_1, X_2, \cdots, X_m$, let $Y_1, Y_2, \cdots, Y_n$ be i.i.d. with distribution F_2 and $D_1, D_2, \cdots, D_n$ from G_2 be the censoring variables associated with $Y_1, Y_2, \cdots, Y_n$ respectively. Y's and D's are independent. We observe $(V_1, \delta_1), (V_2, \delta_2), \cdots, (V_n, \delta_n)$ where $V_j = \min(Y_j \leq D_j)$ and $\delta_j = I[Y_j \leq D_j]$. The two-sample problem is to test $H_0 : F_1 = F_2 = F$ (say) based on the available data (T_i, ϵ_i) and $(V_j, \delta_j), i = 1, \cdots, m, j = 1, 2, \cdots, n$.

(A) Gehan's Test (1965)

A test for the hypothesis of interest suggested by Gehan is an extension of the Mann-Whitney-Wilcoxon's test. It is based on the modified kernel:

$$U_{ij} = \begin{cases} 1 \ \text{if } X_i > Y_j, \text{that is,} \\ \quad t_i > v_j, \delta_j = 1, \text{or} \\ \quad t_i = v_j, \delta_j = 1, \epsilon_i = 0 \\ -1 \ \text{if } \ X_i < Y_j, \ \text{that is,} \\ \quad t_i < v_j, \epsilon_i = 1 \quad \text{or} \\ \quad t_i = v_j, \epsilon_i = 1, \delta_j = 0 \\ 0 \ \text{otherwise.} \end{cases}$$

Then

$$U = \sum_i \sum_j U_{ij}.$$

The distribution of U-statistic is asymptotically normal by the two-sample U-statistic theorem. In order to apply the theorem, first we compute its mean and variance under H_0.

$$E_{H_0}(U_{ij}) = P[T_i > V_j, \delta_j = 1] + P[T_i = V_j, \delta_j = 1, \epsilon_i = 0]$$
$$-P[T_i < V_j, \epsilon_i = 1] - P[T_i = V_j, \delta_j = 0, \epsilon_i = 1]. \qquad (7.3.1)$$

Consider

$$\begin{aligned}
P[T_i > V_j, \delta_j = 1] &= P[T_i > V_j | \delta_j = 1, \epsilon_i = 1] P[\delta_j = 1, \epsilon_i = 1] \\
&+ P[T_i > V_j | \delta_j = 1, \epsilon_i = 0] P[\delta_j = 1, \epsilon_i = 0] \\
&= P[X_i > Y_j] P[\delta_j = 1] P[\epsilon_i = 1] + P[C_i > Y_j] P[\delta_j = 1] P[\epsilon_i = 0] \\
&= P[X_i > Y_j] P[Y_j \le D_j] P[X_i \le C_i] + P[C_i > Y_j] P[Y_j \le D_j] P[X_i > C_i] \\
&= \int_0^\infty \overline{F}_1(u) dF_2(u) \int_0^\infty F_2(u) dG_2(u) \int_0^\infty F_1(u) dG_1(u) \\
&+ \int_0^\infty F_2(u) dG_1(u) \int_0^\infty F_2(u) dG_2(u) \int_0^\infty \overline{F}_1(u) dG_1(u) \\
&= \int_0^\infty \overline{F}(u) dF(u) \int_0^\infty F(u) dG_2(u) \int_0^\infty F(u) dG_1(u) \\
&+ \int_0^\infty F(u) dG_1(u) \int_0^\infty F(u) dG_2(u) \int_0^\infty \overline{F}(u) dG_1(u) \\
&\qquad \text{since} \quad F_1 = F_2 = F. \qquad (7.3.2)
\end{aligned}$$

From (7.3.1) and (7.3.2) it is clear that $E_{H_0}(U)$ involves G_1 and G_2 also. So Gehan has considered the more restrictive hypothesis H_0^* where $H_0^* : F_1 = F_2$ and $G_1 = G_2 = G$ (say). Under $H_0^*, E_{H_0^*}(U_{ij}) = 0$. But the variance of U, even under H_0^* is not free from F and G which are unknown. Mantel (1963) has presented an alternative method for computing U statistic of Gehan and its conditional variance under H_0^* given, what is called the "censoring pattern".

Mantel's Method: Let $Z_1 \le Z_2, \cdots \le Z_{m+n}$ be the combined ordered sample of size $(m+n)$ where Z_i's are either T_i's or V_j's. Then the censoring pattern is

$$(Z_1, \eta_1), (Z_2, \eta_2), \cdots, (Z_{m+n}, \eta_{m+n})$$

where η_i's are either ϵ_i's or δ_j's which are either 1 or 0. We assume that, it is not known whether Z is from sample one or sample two. We only know that m of $m+n$ observations are from sample one. There are $\binom{n+m}{m}$ ways in which m units can be chosen from $(m+n)$ units. Under H_0 each of these orders has the same probability viz. $\frac{1}{\binom{n+m}{m}}$. Thus, the situation is similar to simple random sampling without replacement (SRSWOR).

Define

$$U_{k\ell}^* = U[(Z_k, \eta_k), (Z_\ell, \eta_\ell)]$$

as follows:

$$U_{k\ell}^* = \begin{cases} 1 \text{ if } Z_k > Z_\ell, \eta_\ell = 1 \\ \quad \text{or } Z_k = Z_\ell, \eta_k = 0, \eta_\ell = 1 \\ -1 \text{ if } Z_k < Z_\ell, \eta_k = 1 \\ \quad \text{or } Z_k = Z_\ell, \eta_\ell = 0, \eta_k = 1 \\ 0 \text{ otherwise.} \end{cases}$$

Let

$$U_k^* = \sum_{\substack{\ell=1 \\ \ell \neq k}}^{m+n} U_{k\ell}^*.$$

Then

$$U = \sum_{k=1}^{m+n} U_k^* I(k \in I_1),$$

where I_1 is the set of integers comprising of observations from the first sample. Note that

$$U_{k_1,k_2} = -U_{k_2,k_1} \quad \text{if} \quad k_1, k_2 \in I_1$$

and U is Gehan's statistic.

Now we assume that the censoring pattern is given. Therefore $U_k^*, k = 1, 2, \cdots, m+n$ are known. Under H_0^* we have to choose m of these randomly and add to get the statistic U. That is, we are using SRSWOR and therefore the standard results from SRSWOR are applicable. Hence,

$$V_{H_0^*}(U) = V_{H_0^*}[\sum_{k=1}^{m+n} U_k^* I(k \in I_1)].$$

While applying the results of SRSWOR, it may be noted that we are seeking the variance of the sample total and not of the sample mean.

$$V_{H_0^*}[\sum_{k=1}^{m+n} U_k^* I(k \in I_1)] = m^2 \left[\frac{(m+n-m)}{(m+n)m} V(U_{ij}^*)\right]$$

$$= \frac{mn}{(m+n)(m+n-1)} \sum_{j=1}^{m+n} U_j^{*2}, \quad \text{since} \quad E_{H_0^*}[U_j^*] = 0.$$

The expression for $V_{H_0^*}(U)$, thus obtained is that of conditional variance given the 'censoring pattern'. We carry out the test using the above as an estimator of the unconditional variance.

Illustration 7.1: We shall apply Gehan's test to Byron and Brown's hypothetical data. It is shown in the following table and the corresponding Figure 7.1:

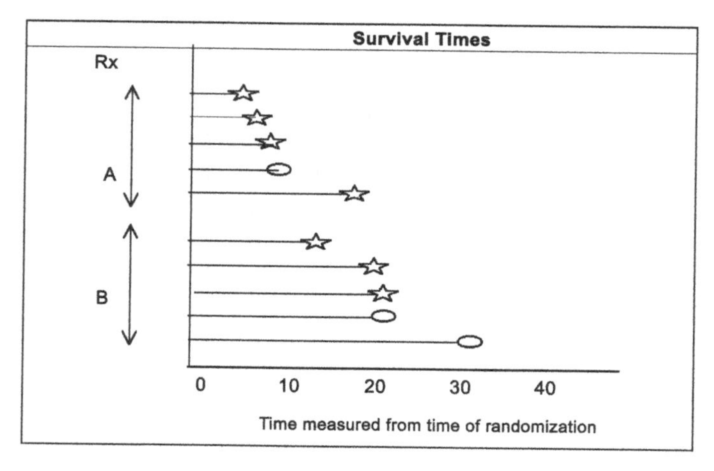

Figure 7.1

Let the treatment 'A' patients be the X observations and treatment 'B' patients be Y observations.

Table 7.1

Byron and Brown's hypothetical trial data

R_X A:	3	5	7	9^+	18
R_X B:	12	19	20	20^+	33^+

Table 7.2

Computations of Gehan's test-statistic for Byron and Brown's hypothetical data

Z	R_X	$\# < Z$	$\# > Z$	U^*
3	A	0	9	−9
5	A	1	8	−7
7	A	2	7	−5
9^+	A	3	0	3
12	B	3	5	−2
18	A	4	4	0
19	B	5	3	2
20	B	6	2	4
20^+	B	7	0	7
33^+	B	7	0	7

$$U = -9 - 7 - 5 + 3 + 0 = -18$$

$$E_{H_0^*}(U) = 0$$
$$Var^*_{H_0}(U) = \frac{(5)(5)(286)}{(10)(9)} = 79.44.$$

Under H_0^*

$$Z_0 = \frac{U}{\sqrt{Var_{H_0}(U)}} = \frac{-18}{8.91} = -2.02$$

so that p-value for one-tailed test is 0.022.

(B) The Mantel-Haenszel Test [(1959), (1963)]

At the heart of this test lies the familiar chi-square statistic for a 2×2 contingency table. Now it is extended to a sequence of 2×2 contingency tables which cannot be combined. This generalisation is then applied to survival data presented as follows at each known survival time (complete observation) by a separate 2×2 contingency table:

	Dead	Alive	Total
I	a	b	$a+b$
II	c	d	$c+d$
Total	$a+c$	$b+d$	$a+b+c+d$

where a: The no. of deaths at 'complete' observation (say t) from sample I
$a+b$: The no. at risk at $t-$(just prior to t) from sample I
$a+c$: The total no. of deaths at $t-$
$a+b+c+d$: The total no. at risk from combined smple at $t-$.

Single 2×2 *Table*

Suppose we have two populations divided by two possible values of characteristics. For example, population I might be cancer patients under a certain treatment and population II comprises of cancer patients under a different treatment. The patients in either group may die within a year or survive beyond a year. The data may be summarised in a 2×2 table.

	Dead	Alive	
Population I	a	b	n_1
Population II	c	d	n_2
	m_1	m_2	n

We use the following notation:

$$P_1 = P(\text{Dead / population I})$$

$$P_2 = P(\text{Dead / population II})$$

so that
$1 - P_1 = P$ (Alive / population I) and
$1 - P_2 = P$ (Alive / population II).
The hypothesis of interest for this two-sample problem is

$$H_0 : P_1 = P_2 = P \quad \text{(say)}.$$

The usual statistic is

$$Z = \frac{\hat{P}_1 - \hat{P}_2}{\sqrt{\hat{P}(1-\hat{P})(\frac{1}{n_1} + \frac{1}{n_2})}} \xrightarrow{a} N(0,1),$$

where

$$\hat{P}_1 = \frac{a}{n_1}, \quad \hat{P}_2 = \frac{c}{n_2}$$

and

$$\hat{P} = \frac{(a+c)}{n_1 + n_2} = \frac{m_1}{n}.$$

Hence

$$Z^2 = \chi_1^2 = \frac{n(ad-bc)^2}{n_1 n_2 m_1 m_2},$$

the usual chi-square for 2×2 contingency table. With continuity correction, the chi-square statistic is

$$\chi_c^2 = \frac{n(|ad-bc| - n/2)^2}{n_1 n_2 m_1 m_2}.$$

Exact hypergeometric distribution: Chi-square distribution of the test statistic as given above is an approximation to the exact distribution. Taking n_1, n_2, m_1 and m_2 as fixed, the distribution under H_0 of the random variable A, which is the entry in the (1,1) cell of the 2×2 table is hypergeometric. Therefore,

$$P[A = a] = \frac{\binom{n-1}{a}\binom{n_2}{m_1 - a}}{\binom{n}{m_1}}, a = 0, 1, 2, \cdots, \min[m_1, n_1].$$

The first two moments of the hypergeometric distribution are:

$$E_{H_0}(A) = \frac{n_1 m_1}{n}, Var_{H_0}(A) = \frac{n_1 n_2 m_1 m_2}{n^2(n-1)}.$$

This gives,

$$\frac{n(ad-bc)^2}{n_1 n_2 m_1 m_2} = \frac{n}{n-1}\left[\frac{a - E_{H_0}(A)}{\sqrt{V_{H_0} A)}}\right]^2$$
$$\xrightarrow{\mathcal{D}} \chi_1^2 \quad \text{as} \quad n \to \infty.$$

Sequence of 2×2 *tables*

Now suppose we have a sequence of 2×2 tables. For example, we might have k hospitals; at each hospital, patients receive either Treatment 1 or Treatment 2 and their responses are observed.

Because there may be differences among hospitals, we do not want to combine all k tables into a single 2×2 table. Based on these $k \times k$ tables, we wish to test the hypothesis:

$$H_0 : P_{11} = P_{12}, \cdots, P_{k1} = P_{k2},$$

where

$P_{i1} = P$ { Dead / Treatment 1, Hospital i },

$P_{i2} = P$ { Dead / Treatment 2, Hospital i }.

Suppose a_i is the number of patients receiving treatment 1 and who died in hospital i. Then Mantel-Haenszel suggested the following statistic:

$$MH = \frac{\sum_{i=1}^{k}(a_i - E_{H_0}(A_i))}{\sqrt{\sum_{i=1}^{k} Var_{H_0}(A_i)}},$$

which after correction for continuity is

$$MH_c = \frac{|\sum_{i=1}^{k}(a_i - E_{H_0}(A_i)| - 1/2}{\sqrt{\sum_{1}^{k} Var_{H_0}(A_i)}}.$$

If the tables are independent, then $MH \xrightarrow{a} N(0,1)$ either when k is fixed and $n_i \to \infty$ or when $k \to \infty$ and A_i's are identically distributed.

Application of MH statistics in survival analysis

Recall that $(Z_{(1)}, \eta_{(1)}), \cdots, (Z_{(n_1+n_2)}, \eta_{(n_1+n_2)})$ is a combined ordered sample, where $\eta(\cdot)$ is censoring indicator of $Z(\cdot)$. It is assumed that the censoring pattern is known. Construct a 2×2 table for each of the uncensored time point. Compute the MH statistic for this sequence of tables to test $H_0 : F_1 = F_2$.

The following figure illustrates the MH sequence of 2×2 tables:

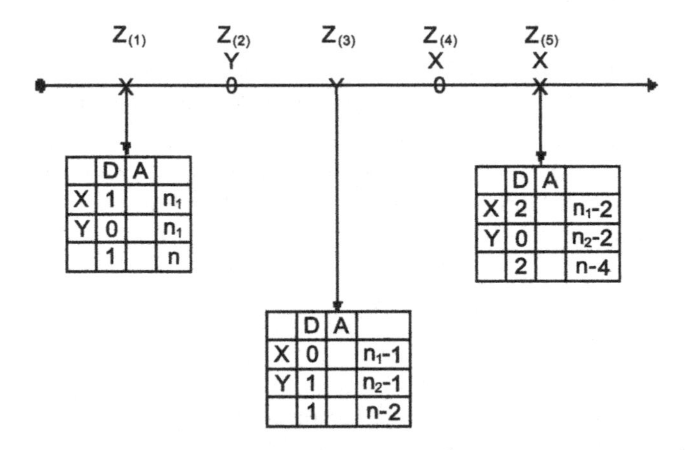

Figure 7.2

It may be noted that the tables are not independent.

Illustration 7.2: We shall illustrate computation of MH statistic for Byron and Brown's hypothetical data. Table 7.3 shows the computations of MH statistic. The column labelled Z contains the uncensored ordered observations. The next four columns labelled n, m_1, n_1 and 'a', construct the 2×2 tables:

a		n_1
m_1		n

The next column is $E_{H_0}(A) = \frac{n_1 m_1}{n}$. The product of the entries in the remaining two columns, labelled $\frac{m_1(n-m_1)}{(n-1)}$ and $(\frac{n_1}{n})(1 - \frac{n_1}{n})$ is $Var_{H_0}(A)$.

Table 7.3
Computations for the Mantel-Haenszel statistic in Byron and Brown's hypothetical data

Z	n	m_1	n_1	a	$E_{H_0}(A)$	$a - E_{H_0}(A)$	$\frac{m_1(n-m_1)}{n-1}$	$\frac{n_1}{n}(1-\frac{n_1}{n})$
3	10	1	5	1	0.50	0.50	1	0.2500
5	9	1	4	1	0.44	0.56	1	0.2469
7	8	1	3	1	0.38	0.62	1	0.2344
12	6	1	1	0	0.17	-0.17	1	0.1389
18	5	1	1	1	0.20	0.80	1	0.1600
19	4	1	0	0	0	0	1	0
20	3	1	0	0	0	0	1	0

For example, (i) first 2×2 table is

1	4	5
0	5	5
1	9	10

Therefore,

$$E_{H_0}(A) = \frac{1 \times 5}{10} = 0.50.$$

(ii) 2×2 table corresponding to Z-value 12 is

0	1	1
1	4	5
1	5	6

Therefore,

$$E_{H_0}(A) = \frac{1 \times 1}{6} = 0.17.$$

$$MH = \frac{\sum_i (a_i - E_{H_0}(A_i))}{\sqrt{\sum_i \frac{m_{1i}(n_i - m_{1i})}{(n_i - 1)} \times \frac{n_{1i}}{n_i}(1 - \frac{n_{1i}}{n_i})}}$$
$$= \frac{\sum_i (a_i - E_{H_0}(A_i))}{\sqrt{\sum_i V_{H_0}(A_i)}}.$$

$$MH = \frac{2.31}{\sqrt{1.03}} = 2.24.$$

P-value $= 0.012$ (one-tailed test).

$$MH_c = \frac{2.31 - 0.5}{\sqrt{1.03}} = 1.76.$$

p-value $= 0.039$ (one-tailed test).

Today, Kaplan-Meier estimation of survival probabilities and the Mantel-Haenszel test for comparison of two survival curves are the two procedures which remain even after a long span of years, the most widely used techniques in the clinician's tool box for survival analysis.

(C) Tarone-Ware class of tests: (Biometrika, 1977).

After constructing a 2×2 table for each uncensored observation, Tarone and Ware suggest weighing each table to form the staistic,

$$\sum_{i=1}^{k} W_i[a_i - E_{H_0}(A_i)] = \sum_{i=1}^{k} W_i[a_i - \frac{m_{1i}n_{1i}}{n_i}]. \qquad (*)$$

The variance of the statistic (*) is given by

$$\sum_{i=1}^{k} W_i^2 Var_{H_0}(A_i) = \sum_{i=1}^{k} W_i^2 \left[\frac{m_{1i}(n_i - m_{1i})}{(n_i - 1)}\right] \times \left[(\frac{n_{1i}}{n_i})(1 - \frac{n_{1i}}{n_i})\right]. \qquad (**)$$

There are three important special cases:

(i) $W_i = 1$ gives MH statistic.

(ii) $W_i = n_i$ gives the Gehan statistics but $\hat{Var}_{TW}(U)$ given by (**) is not the same as $Var_{H_0^*}(U)$. However, they are asymptotically equivalent under H_0.

(iii) $W_i = \sqrt{n_i}$ is suggested by Tarone and Ware.

It may be noted that Gehan's statistic puts more weight on the small observations, while MH statistic puts equal weight on each observation. Tarone and Ware's suggestion is intermediate between the two. They claim that the weights $W_i = \sqrt{n_i}$ have greater efficiency over a range of alternatives.

Illustration 7.3. We again consider Byron and Brown's hypothetical data. Referring to the table of computations for MH statistic for Byron and

Brown's hypothetical data, we get

$$\sum_{i=1}^{k} n_i(a_i - E_0(A_i)) = 17.98.$$

This is what we got for Gehan's statistic except for the sign and roundoff error. Also

$$\hat{Var}_{TW}(U) = \left[n_i^2 \left[\frac{m_{1i}(n_i - m_i)}{(n_i - 1)}\right]\right] \left[\left(\frac{n_{1i}}{n_i}\right)\left(1 - \frac{n_{1i}}{n_i}\right)\right] = 69.2439.$$

$$Var_{H_0^*}(U) = 79.44$$

which gives

$$\sqrt{\hat{Var}_{TW}(U)} = 8.3213$$

and

$$\sqrt{Var_{H_0^*}(U)} = 8.9129.$$

(D) Log-Rank Test

The log-rank test so named by Peto and Peto (1972) is another rank test for comparing two distributions on the basis of randomly right censored data.

Let $\tau_1 < \tau_2 \cdots < \tau_g$ be the distinct ordered failure epochs from the combined sample.

d_j = The number of deaths at τ_j, (for complete data with no ties $d_j = 1$ for every j).

r_j = The number at risk at τ_j.

r_{ij} = The number at risk at τ_j from group i for $i = 1, 2$.

The log-rank test compares the observed and expected (under H_0) number of deaths in group I.

$$
\begin{aligned}
E &= \text{Expected (conditional under } H_0\text{) number of deaths in sample I} \\
&= \sum_{j=1}^{g} d_j \frac{r_{1j}}{r_j} \\
V &= \sum_{j=1}^{g} d_j \frac{r_{1j} r_{2j}}{r_j^2} \\
O &= \text{observed number deaths in sample (group) I} \\
Z &= \frac{0 - E}{\sqrt{V}} \xrightarrow{a} N(0, 1).
\end{aligned}
$$

Illustration 7.4: The following table shows failure time of two machines, new and old.

	Failure times (day)
New machine	250, 476^+, 355, 200, 355^+
Old machine	191, 563, 242, 285, 16, 16, 16, 257, 16

(+ indicates censored times).

Test whether the new machine is more reliable than the old one by using log-rank test.

Solution: $\chi_1^2 = 3.7$

p-value for two sided test = 0.055

p-value for one sided test = 0.0275.

R-commands are given in the appendix of this chapter.

Conclusion: The new machine is more reliable than the old one by using log-rank test at 5% level of significance.

Remark: The two sample tests described in this chapter are applicable when one survival curve dominates the other. In case of crossing of survival curves these methods should be used with caution. For example, the MH test is known to be the most sensitive method of detecting departures from the null hypothesis $\overline{F}_1(x) = \overline{F}_2(x)$ when the true relationship is of the form

$$\overline{F}_2(x) = [\overline{F}_1(x)]^k.$$

However, if two survival curves were to cross each other near their mid-points, the total number of observed deaths in each group would approximately be equal to the expected number and the null hypothesis, though clearly false would not be rejected by the MH test.

An example of practical situation where we come across crossing survival curves is visualised by plotting groupwise survival curves for the data of lymphocytic, non-Hodgkins lymphoma of exercise 5.2.

In such a situation, weighted tests are better than the MH test. The reason for this will be clear if we consider the alternative of logistic functional relationship. Specifically

$$S_1(t) = [1 + \exp(\lambda_1 t)]^{-1}, t \geq 0, \lambda_1 > 0$$

$$S_2(t) = [1 + \exp(\lambda_2 t)]^{-1}, t \geq 0, \lambda_2 > 0$$

then

$$S_2(t) = \left\{1 + \left[\frac{1 - S_1(t)}{S_1(t)}\right]^k\right\}^{-1}, k = \frac{\lambda_2}{\lambda_1}.$$

In this case, for any value of $k \neq 1$, the survival curves will cross when $S_1(t) = S_2(t) = 0.5$. For $0 < k < 1$, $S_1(t)$ will dominate $S_2(t)$ when $S_1(t) > 0.5$ and $S_2(t)$ will dominate $S_1(t)$ when $S_1(t) < 0.5$. The converse relationship holds when $k > 1$. Since the MH-test gives equal weights to both negative and positive differences between observed and expected frequencies as noted earlier, it is likely to produce a non-significant result. The Gehan and other tests which give greater weight to differences at the values of $S(t)$ above 0.5 will not be so affected. They are more likely to generate a significant statistic under the logistic and similar alternative. Another kind of relationship between survival curves appears in data reported by Fleming et al. (1980) on cumulative rates of disease progression over time in two different categories of patients. The rates appear almost identical for the first half of study period and then suddenly diverge. In such a situation the MH test is more likely to show a significant result than the tests that weight earlier differences more heavily than the later ones.

Exercises

Exercise 7.1: Ten female patients with breast cancer are randomised to receive either a treatment or no treatment, say control, after a radical mastectomy. At the end of two years, the following times to relapse in months are recorded:

Treatment (group 1): 23, 16+, 18+, 20+, 24+

Control (group 2): 15, 18,19, 19,20
The testing problem is:
H_0: Treatment and control are equally effective
H_1: Treatment is more efficient than no treatment.
Carry out test of the hypothesis using:

1. Gehan test
2. Mantel-Haenszel test
3. Tarone-Ware test with optimum weights suggested by Tarone-Ware
4. Log-rank test of Peto and Peto.

Exercise 7.2: The following data arose from an experiment investigating motion sickness at sea level reported by Burns (1984), and also by Altman (1991, pp. 368-371). The study subjected individuals to vertical motion for two hours, recording the time until each subject vomited or asked to withdraw. Those in group I were given frequency 0.167 Hz and acceleration / 0.1111 g, while those in group II were subjected to a double dose (0.3333 Hz, and 0.2222 g). An asterisk denotes withdrawal or "successful" completion.
Group I: $33, 50, 50^*, 51, 66^*, 82, 92, 120^*, 120^*, 120^*, 120^*,$
$120^*, 120^*, 120^*, 120^*, 120^*, 120^*, 120^*, 120^*, 120^*, 120^*.$
Group II: $5, 6^*, 11, 11, 13, 24, 63, 65, 69, 69, 79, 82, 82, 102, 115, 120^*,$
$120^*, 120^*, 120^*, 120^*, 120^*, 120^*, 120^*, 120^*, 120^*, 120^*, 120^*, 120^*.$

Analyse the data using log-rank test (Peto and Peto).

Exercise 7.3: Veteran is a standard survival data set of lung cancer study (Randomised clinical trial of two treatments). Treatment 1 is standard treatment and treatment 2 is test treatment. Extract the resident data set from survival package with commands:
library(survival) # Attach library survival;
data(veteran);
veteran;
Construct a plot survival curves for two treatments. Is it advisable to carry out the log-rank test to compare survival experience of the two treatments? Justify your answer.

Exercise 7.4: Use the Leukemia (AML) data of illustration 5.4. Formulate the null and alternative hypothesis for the comparison of the survival experience of the two groups. Apply log-rank test of Peto and Peto. Write your conclusion.

Appendix

Two-sample tests.

R-commands implement the two-sample family of tests for equality of survival distributions of Harrington and Fleming (1982), with weights rho. With 'rho = 1' this is the log-rank test of Peto and Peto.

```
> time<-c(250,476,355,200,355,191,563,242,285,16,16,16,257,16);

> length(time);
[1] 14

> status<-c(1,0,1,1,0, rep(1,9));

> length(status);
[1] 14

> gr<-c(rep(1,5),rep(2,9));

> library(survival);

> two.sample<-data.frame (time, status, gr);

> survdiff(Surv(time,status)~ gr,rho=1);


Call:
survdiff(formula = Surv(time, status) ~ gr, rho = 1)

     N Observed Expected (O-E)^2/E (O-E)^2/V
gr=1 5     1.43     3.36     1.108      3.18
gr=2 9     6.29     4.36     0.854      3.18

 Chisq= 3.2  on 1 degrees of freedom, p= 0.0747
```

Note: p-value is given for two-tailed test. So p-value for one-tailed test is 0.0275.

Chapter 8

Proportional Hazards and Other Regression Models

8.1 Introduction

In this chapter, we shall study certain models which incorporate the effects of covariates or explanatory variables on the distributions of the lifetimes. Within these models we will be able to test whether the covariates affect the lifetimes significantly or not.

Covariates: These are the characteristics or features of the experimental units which are thought to affect the lifetimes of individuals. The following are some examples of covariates:

(i) *Treatments*: In simple comparison of two treatments, say a "new" treatment with a "control" or "standard" treatment, we consider a binary covariate Z defined as:

$$Z = \begin{cases} 1, & \text{if an individual receives "new" treatment.} \\ 0, & \text{if an individual receives "control" or "standard" treatment.} \end{cases}$$

If a treatment is specified by the dose, then the corresponding covariate is the value of the dose (or log dose).

(ii) *Intrinsic Properties*: Explanatory variables or covariates, measuring intrinsic properties of the individual, include (in medical context) such variables as sex, age on entry in the medical trial and variables describing medical history before admission to the study.

(iii) *Exogeneous Variables*: This type of covariates exhibits environmental features of the problem, for example, grouping of individuals according to observers or apparatus, month in which the experiment was carried out, etc.

The covariates could be constant over time or dependent on time. For example,

(i) Suppose that a treatment is applied at time $t_0 > 0$. Then one can incorporate a time dependent binary covariate ($Z(t)$) defined as

$$Z(t) = \begin{cases} 0 \text{ if } t < t_0 \\ 1 \text{ if } t \geq t_0. \end{cases}$$

(ii) In some industrial applications, a time varying stress may be applied. So the covariate process will be the entire history of the stress process.

We shall not discuss time dependent covariates further.

Model Formulation: Suppose that for every individual there is defined a $q \times 1$ vector $\underline{Z}$ of covariates. It is often convenient to define $\underline{Z}$ such that $\underline{Z} = \underline{0}$ corresponds to some meaningful "standard" conditions. Such models are developed in two parts:

(a) A model for the distribution of lifetime when $\underline{Z} = \underline{0}$ which may be called the baseline model and

(b) A representation (link function) of the changes introduced by the non-zero vector $\underline{Z}$.

The baseline model (when $\underline{Z} = \underline{0}$) could be parametric or non-parametric and link function is usually parametric. Accordingly the combined regression model will be parametric or semiparametric.

Proportional Hazards Model (PH Model):

This is the most cited and used model which is introduced by Cox (1972) in his path breaking paper. The simplest form of proportional hazards model is:

$$h(t, \underline{Z}) = h_0(t)\psi(\underline{Z}), \tag{8.1.1}$$

where $h_0(t)$ is the baseline failure rate and $\psi(\underline{Z})$ is the link function bringing in the covariates. It satisfies $\psi(\underline{0}) = 1$ and $\psi(\underline{Z}) \geq 0$ for all $\underline{Z}$. Note that the failure rate function $h(t, \underline{z})$ is a function of time t as well as the covariate values $\underline{z}$.

The following two parametric link functions are commonly used:

(i) $\psi(\underline{Z}; \underline{\beta}) = \exp(\underline{\beta}'\underline{Z})$: log linear form.

(ii) $\psi(\underline{Z}; \underline{\beta}) = 1 + \underline{\beta}'\underline{Z}$: linear form.

We shall consider the base line hazard rate, $h_0(t)$, as completely unknown and the covariates as fixed quantities, thus leading to the semiparametric PH model.

8.2 Complete Data

Let $t_1, t_2, \cdots, t_n$ be the observations and $\tau_1 < \tau_2 < \cdots < \tau_n$ denote ordered failure times of the n individuals. Let ℓ_j be the label of the subject which fails at τ_j. Thus $\ell_j = i$ iff $t_i = \tau_j$. Let $R(\tau_j)$ be the risk set at time point τ_j. Then $R(\tau_j) = \{i; t_i \geq \tau_j\}$. For example consider the following data:

Individual i	failure time t_i	j	$R(\tau_j)$
1	20	3	1
2	10	1	1, 2, 3
3	15	2	1, 3

The following figure illustrates these concepts.

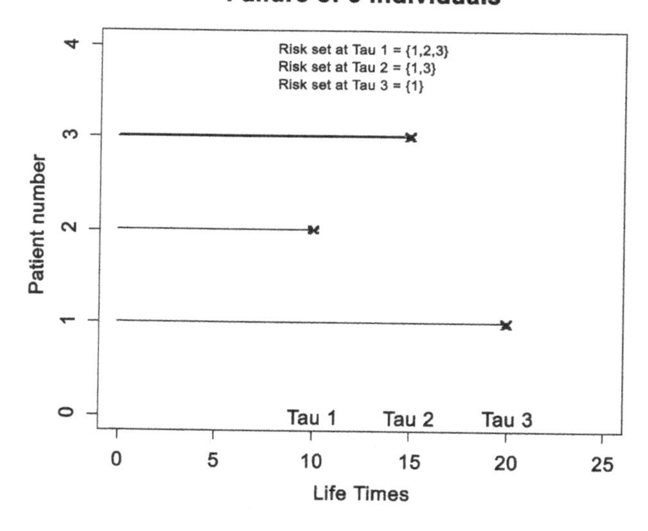

Figure 8.1

Likelihood Function

The basic principle of the derivation of the likelihood is as follows. The $\{\tau_j\}$ and $\{\ell_j\}$ are jointly equivalent to the original data, namely the unordered failure times t_i. In the absence of knowledge of $h_0(t)$, the τ_j can provide little or no information about $\underline{\beta}$ as their distribution depends heavily on $h_0(t)$. As an extreme example, $h_0(t)$ might be identically zero except in the small neighbourhood of τ_j. Therefore we focus our attention on the ℓ_j's. In the present case, their conditional joint distribution $p(i_1, \cdots, i_n)$ over the set of all possible permutations can be derived explicitly. The conditional probability that $\ell_j = i$ given the entire history, $H_j = \{\tau_1, \cdots, \tau_j, i_1, i_2, \cdots, i_{j-1}\}$, upto the j-th failure time τ_j can be written as

$$P_j[\ell_j = i|H_j] = \text{The conditional probability that } i \text{ fails at } \tau_j \text{ given that one individual from } R(\tau_j) \text{ fails at } \tau_j$$

$$= \frac{h_i(\tau_j)}{\sum_{k \in R(\tau_j)} h_k(\tau_j)} = \frac{\psi(i)}{\sum_{k \in R(\tau_j)} \psi(k)}. \tag{8.2.1}$$

The baseline hazard function getting cancelled because of the multiplicative form of the model. For notational convenience $\psi(k)$ here denotes $\psi(\underline{Z}_k, \underline{\beta})$ which is the multiplier for the k-th subject. Although (8.2.1) was derived as the conditional probability that $\ell_j = i$ given the entire history H_j, in fact, it is independent of $\tau_1, \tau_2, \cdots, \tau_j$. It therefore, equals $p_j(i/i_1, \cdots, i_{j-1})$. Thus

$$P_j(\ell_j = i|H_j) = P_j(i/i_1, \cdots, i_{j-1}) = \frac{\psi(i)}{\sum_{k \in R(\tau_j)} \psi(k)}.$$

The joint distribution $P(i_1, i_2, \cdots, i_n)$ can therefore be obtained by the usual chain rule for conditional probabilities as follows:

$$P(i_1, i_2, \cdots, i_n) = \prod_{j=1}^{n} P_j(\ell_j/i_1, \cdots, i_{j-1}) = \prod_{j=1}^{n} \frac{\psi(\ell_j)}{\sum_{k \in R(\tau_j)} \psi(k)}. \tag{8.2.2}$$

As an example, consider the configuration of Figure 8.1,

$$P[2, 3, 1] = \frac{\psi(2)}{\psi(1) + \psi(2) + \psi(3)} \times \frac{\psi(3)}{\psi(1) + \psi(3)} \times \frac{\psi(1)}{\psi(1)}.$$

Equation (8.2.2) is called the *partial likelihood* of $(i_1, i_2, \cdots, i_n)$.
Illustration 8.1

A set of $n = 3$ light bulbs are placed on test. The first and second bulbs are 100 watt bulbs and the third bulb is 60 watt bulb. A single $(q = 1)$ covariate Z assumes the value 0 for the 60-watt bulb and 1 for the 100-watt bulbs. The purpose of the test is to determine if the wattage has any influence on the survival distribution of the bulbs. The baseline distribution is unknown and unspecified, so there is only one parameter in the proportional hazards model, the regression coefficient β, which needs to be estimated. This small data set is used for illustrative purposes only. We would obviously need to collect more than three data points to detect the significance of difference between the two wattages. Let $t_1 = 80, t_2 = 20$ and $t_3 = 50$ denote the lifetimes of the three bulbs, then we have from Figure 8.2 $\tau_1 = t_2 = 20, \tau_2 = t_3 = 50, \tau_3 = t_1 = 80, z_1 = 1, z_2 = 1, z_3 = 0$.

The mass function for the observed ordered vector [2, 3, 1] will be determined by finding the conditional probabilities associated with the ranks. For example, assume that a failure has just occurred at time $\tau_2 = 50$ and history up to time 50, (that is, bulb labeled 2 has failed at time 20) is known. So the bulb that fails at 50 is either bulb 1 or bulb 3. The conditional probability that the bulb failing at time 50 is bulb 1 is

$$P[\ell_2 = 1 | \tau_1 = 20, \tau_2 = 50, \ell_1 = 2] = \frac{\psi(1)}{\psi(1) + \psi(3)}$$

$$= \frac{e^\beta}{e^\beta + 1}, \text{ assuming the log linear form of the link function.}$$

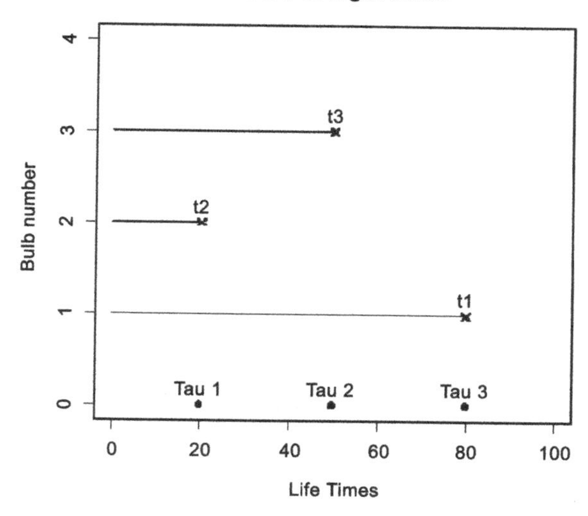

Figure 8.2

Note that the baseline hazard function has dropped out of this expression, so the probability will be the same regardless of the choice of $h_0(t)$. Furthermore note that τ_1 and τ_2 are not used in the calculation. Hence one can write

$$P[\ell_2 = 1|\tau_1 = 20, \tau_2 = 50, \ell_1 = 2] = P[\ell_2 = 1/\ell_1 = 2].$$

Thus,

$$\begin{aligned}
L(\beta) &= P(2,3,1) = p(2)p(3/2)p(1/2,3) \\
&= \frac{\psi(2)}{\psi(1)+\psi(2)+\psi(3)} \times \frac{\psi(3)}{\psi(1)+\psi(3)} \times \frac{\psi(1)}{\psi(1)} \\
&= \frac{e^{\beta}}{e^{\beta}+e^{\beta}+1} \cdot \frac{1}{e^{\beta}+1} = \frac{e^{\beta}}{(2e^{\beta}+1)(e^{\beta}+1)}
\end{aligned}$$

and

$$\log L(\beta) = \beta - \log(2e^{\beta}+1) - \log(e^{\beta}+1).$$

The score function is

$$\frac{\delta}{\delta\beta} \log L(\beta) = 1 - \frac{2e^{\beta}}{2e^{\beta}+1} - \frac{e^{\beta}}{e^{\beta}+1}.$$

Solving the maximum likelihood equation using numerical methods, we obtain $\beta = -0.347$ indicating that there is lower risk for 100-watt bulbs than for 60-watt bulbs. More specifically, the hazard function for 100-watt bulbs is $e^{\hat{\beta}} = e^{-0.347} = 0.706$ times that of the baseline hazard function for 60-watt bulbs, regardless of the form of baseline distribution. This we shall denote as hazard ratio (HR).

To test the statistical significance of the regression coefficient β we find

$$\frac{-\delta^2 \log L(\beta)}{\delta\beta^2} = \frac{2e^{\beta}}{(2e^{\beta}+1)^2} + \frac{e^{\beta}}{(e^{\beta}+1)^2}.$$

This expression is evaluated at $\beta = \hat{\beta}$, to get the sample information. For the problem under consideration it is 0.485 and the reciprocal 2.06 is the asymptotic estimate of the variance. Thus the asymptotic estimate of the standard deviation is 1.44. Hence based on Wald statistic we conclude that β is not significantly different from zero.

The result is not surprising, as the sample size is very small. Note again that $n = 3$ is taken only for the simplicity of the exposition. The theory actually holds only for large sample sizes. In addition it may be noted that the order and not the magnitude of the failure times is used to find $\hat{\beta}$.

This means, for example, the third bulb could have failed anywhere in the interval (20, 80) without affecting the estimate.

8.3 Censored Data

Suppose that, there are d observed failures from the sample of size n and let the ordered observed failure times be $\tau_1 < \tau_2 < \cdots < \tau_d$. As before let $\ell_j = i$ if the subject i fails at τ_j and let $\mathcal{R}(\tau_j) = \{i; t_i \geq \tau_j\}$ be the corresponding risk set. Equation (8.2.1) follows exactly as before, where H_j now includes the censorings in $(0, \tau_j)$ as well as information regarding the failures and the combination of these conditional probabilities gives the overall partial likelihood:

$$L = \prod_{j=1}^{d} \frac{\psi(\ell_j)}{\sum\limits_{k \in R(\tau_j)} \psi(k)} = \prod_{i \in D} \frac{\psi(i)}{\sum\limits_{k \in R_i} \psi(k)} \tag{8.3.1}$$

where D is the set of complete observations, $R_i = R(t_i)$ and t_i's are the unordered failure times.

It may be noted that we have omitted the terms corresponding to the censored individuals from each risk set. As long as censoring mechanism itself does not depend on $\underline{\beta}$, such terms can be ignored for the purpose of the likelihood inference about $\underline{\beta}$.

Alternatively (8.3.1) can be derived as the sum of probabilities (8.2.2) which are consistent with the observed pattern of failures and censorings.

Illustration 8.2

Figure 8.3 summarises the information regarding the failure of four individuals with censoring.

From this figure

$R_3 = \{1, 2, 3, 4\} = R(\tau_1)$
$R_1 = \{1, 2\} = R(\tau_2)$ (3 has failed and 4 censored before τ_2)
$R_2 = \{2\} = R(\tau_3)$ (1 fails before τ_3).
The partial likelihood is

$$L = \frac{\psi(3)}{\psi(1) + \psi(2) + \psi(3) + \psi(4)} \times \frac{\psi(1)}{\psi(1) + \psi(2)} \times \frac{\psi(2)}{\psi(2)}. \tag{8.3.2}$$

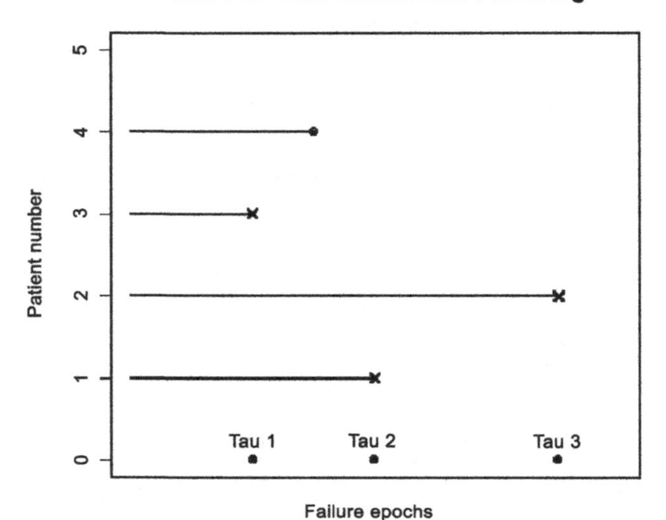

Figure 8.3

x-failure: o - censoring. Failure instants τ_1, τ_2, τ_3.

For the general case the log likelihood is

$$\log(L) = \sum_{i \in D} [\log \psi(i) - \log \sum_{k \in R_i} \psi(k)] = \sum_{i \in D} L_i \ , \ \text{say}.$$

Assume that $\psi(i)$ possesses first and second derivatives with respect to $\underline{\beta}$ for all i. Then

$$\frac{\delta L_i}{\delta \beta_r} = \frac{\psi_r(i)}{\psi(i)} - \frac{\sum\limits_{k \in R_i} \psi_r(k)}{\sum\limits_{k \in R_i} \psi(k)} \tag{8.3.3}$$

where

$$\psi_r(i) = \frac{\delta}{\delta \beta_r} \psi(i), \quad r = 1, 2, \cdots, q.$$

Let

$$\psi_{rs}(i) = \frac{\delta^2}{\delta \beta_r \delta \beta_s} \psi(i) \quad r = 1, 2, \cdots, q, s = 1, 2, \cdots, q.$$

Then

$$\frac{\delta^2 L_i}{\delta\beta_r\delta\beta_s} = \frac{\psi_{rs}(i)}{\psi(i)} - \frac{\psi_r(i)\psi_s(i)}{[\psi(i)]^2} - \frac{\sum\limits_{k\in R_i}\psi_{rs}(k)}{\sum\limits_{k\in R_i}\psi(k)} + \frac{\sum\limits_{k\in R_i}\psi_r(k)\sum\limits_{k\in R_i}\psi_s(k)}{[\sum\limits_{k\in R_i}\psi(k)]^2}. \tag{8.3.4}$$

For the log linear form:

$$\frac{\delta L_i}{\delta\beta_r} = z_{ir} - \frac{\sum\limits_{k\in R_i} z_{kr}\exp[\underline{\beta}'\underline{Z}_k]}{\sum\limits_{k\in R_i}\exp[\underline{\beta}'\underline{Z}_k]},$$

where z_{ir} = value of the r-th covariate for i-th subject.

The last expression is the value of the explanatory variable on the failed subject minus weighted average of the same variable over the corresponding risk set.

$$\frac{\delta L_i}{\delta\beta_r} = z_{ir} - A_{ir}(\underline{\beta}) \text{ say.} \tag{8.3.5}$$

Also

$$\begin{aligned} -\frac{\delta^2 L_i}{\delta\beta_r\delta\beta s} &= -\frac{\sum\limits_{k\in R_i} z_{kr}z_{ks}\exp[\underline{\beta}'\underline{Z}_k]}{\sum\limits_{k\in R_i}\exp[\underline{\beta}'\underline{Z}_k]} \\ &\quad + \frac{\sum\limits_{k\in R_i} z_{ks}\exp[\underline{\beta}'\underline{Z}_k](\sum\limits_{k\in R_i} z_{kr}\exp[\underline{\beta}'\underline{Z}_k])}{[\sum\limits_{k\in R_i}\exp[\underline{\beta}'\underline{Z}_k]]^2} \\ &= C_{irs}(\underline{\beta}), \text{ say.} \end{aligned}$$

The score function is

$$U_r(\underline{\beta}) = \sum_{i\in D}\frac{\delta L_i}{\delta\beta_r} = \sum_{i\in D}(z_{ir} - A_{ir}(\underline{\beta})) \tag{8.3.6}$$

and the sample information matrix is

$$i(\underline{\beta}) = ((\sum_{i\in D} C_{irs}(\underline{\beta}))). \tag{8.3.7}$$

So that Fisher information is

$$E(i(\underline{\beta})) = I(\underline{\beta}). \tag{8.3.8}$$

The likelihood based inference may now be carried out. To estimate $\underline{\beta}$, we solve $U(\underline{\beta}) = \underline{0}$ (q equations in q unknowns). To test the hypothesis $\underline{\beta} = \underline{0}$, the three methods of analysis, namely; the likelihood ratio test, the score test and direct use of maximum likelihood estimates, are applicable.

Exact test for $\underline{\beta} = \underline{0}$:

In the absence of censoring or if the censoring mechanism is independent of the explanatory variables, an exact test of null hypothesis $\underline{\beta} = \underline{0}$ can be obtained by referring the score statistic:

$$U(\underline{0}) = \sum_{i \in D} [\underline{Z}_i - A_i(\underline{0})] \tag{8.3.9}$$

to its permutation distribution. This is the distribution of $U(\underline{0})$ generated when the ordered failure times $\tau_1, \tau_2, \cdots, \tau_d$ and the sizes of the corresponding risk sets $r_1, r_2, \cdots, r_d$ are taken as fixed and the explanatory variables are permuted randomly among the n subjects. This is precisely the conditional distribution of $U(\underline{0})$ given $r_1, r_2, \cdots, r_d$ under the full model. Under the permutation distribution

$$E(U(\underline{0})) = \underline{0} \tag{8.3.10}$$

and the covariance matrix is

$$\tilde{I}(\underline{0}) = \frac{1}{(n-1)} (\sum_{i=1}^{n} (Z_i - \overline{Z})(Z_i - \overline{Z})') \times (\sum_{i=1}^{n} q_i^2), \tag{8.3.11}$$

where

$$\overline{Z} = \sum \frac{Z_i}{n}, \quad q_i = \delta_i - \sum_{j: \tau_j \leq t_i} \frac{1}{r_j}$$

and $\delta_i = 0$ or 1 according as the i-th individual is censored or uncensored. This covariance matrix differs from $I(\underline{0})$.

The Problem of Ties

When there are several failures at the same epoch, each is assumed to contribute the same term to the likelihood function. Consequently all the items with tied failure times are included in the risk set at the time at which multiple failures are observed. This works well when there are only a few ties in the data set.

Many of the softwares use this approximate procedure.

For example, if two items $i = 1, 2$ are observed to fail at τ, from items $i = 1, 2, 3, 4$ at risk, the contribution to the likelihood from τ would be

$$\frac{2\psi(1)\psi(2)}{[\psi(1) + \psi(2) + \psi(3) + \psi(4)]^2}$$

and more generally, if m failures of individuals labelled $i_1, i_2, \cdots, i_m$ are observed at τ then the contribution to the likelihood is

$$\frac{m! \prod_{\ell=1}^{m} \psi(i_\ell)}{[\sum_{k \in R(\tau)} \psi(k)]^m}.$$

This approximation gives tractable log likelihoods for the log linear model.
Illustration 8.3

We analyse the remission time data (Freireich et al. 1963) through the PH model using the R package. These data involve two groups of leukemia patients, with 21 patients in each group. Group 1 is the treatment group and group 2 is the placebo group. The data set also contains the variable log WBC, which is a well-known prognostic indicator of survival for leukemia patients. For this example, the basic question of interest concerns comparing the survival experiences of the two groups adjusting for the possible confounding and or interaction effect of log WBC.

Table 8.1

Leukemia remission data

Sr. No.	Group I		Group II	
	t(weeks)	log WBC	t(weeks)	log WBC
1.	6	2.31	1	2.80
2.	6	4.06	1	5.00
3.	6	3.28	2	4.91
4.	7	4.43	2	4.48
5.	10	2.96	3	4.01
6.	13	2.88	4	4.36
7.	16	3.60	4	2.42
8.	22	2.32	5	3.49
9.	23	2.57	5	3.97
10.	6+	3.20	8	3.52

Table 8.1 (Cont'd)

Sr. No.	Group I		Group II	
	t(weeks)	log WBC	t(weeks)	log WBC
11.	9+	2.80	8	3.05
12.	10+	2.70	8	2.32
13.	11+	2.60	8	3.26
14.	17+	2.16	11	3.49
15.	19+	2.05	11	2.12
16.	20+	2.01	12	1.50
17.	25+	1.78	12	3.06
18.	32+	2.20	15	2.30
19.	32+	2.53	17	2.95
20.	34+	1.47	22	2.73
21.	35+	1.45	23	1.97

Here we have a problem involving two explanatory variables as predictors of the survival time (t). The first explanatory variable is labelled as group and second explanatory variable as log WBC. The variable group is of primary interest. The variable log WBC is a secondary variable that is a possible confounder or effect modifier. We are also interested in the possible interaction effect of log WBC on group. So we include as the third variable group $\times$ log WBC.

For this data set, the computer results from fitting three different PH models are presented below. The software used is R. There are several packages like R, e.g. SPSS, SAS, BMDP etc. which provide analysis by PH model. All these packages provide the same information but possibly in different format.

We now describe how to use the computer printout to evaluate the possible effect of treatment on remission time adjusted for the potential confounding and interaction effects of the covariate log WBC.

Output from R

Model 1:

We consider pH model with single regressor 'group'. The following tables give the output:
$n = 42$

	Coef.	exp(coef.)	se(coef)	Z	p
group	-1.57	0.208	0.412	-3.81	0.00014

	exp(coef)	exp(-coef)	lower 0.95	upper 0.95
group	0.208	4.82	0.0925	0.466

R-square = 0.322 (max. possible = 0.988).
Likelihood ratio test = 16.4 on 1 df, $p = 5.26e^{-5}$
Wald test = 14.5 on 1 df., $p = 0.000138$
Score (log rank) test = 17.5 on 1 df., $p = 3.28e^{-5}$

The coef is the logarithm of the estimated hazard ratio between the two groups, which for convenience is also given as the actual hazards ratio, exp (coef). The next line gives the inverted hazards ratio (swapping the groups) and confidence intervals for the hazards ratio. Finally, three (asymptotically equivalent) tests for testing significance of the group effect are given. Wald test is equivalent to the Z test based on estimated coefficient divided by its standard error, whereas the score test is equivalent to the log rank (MH) test.

It may be noted that except for hazards ratio the output is similar to the output in any regression analysis.

Model 2:
$n = 42$

	coef	exp(coef)	se (coef)	Z	p
group	−1.39	0.25	0.425	−3.26	1.1^{-03}
log WBC	1.69	5.42	0.336	5.03	$4.8e^{-07}$

	exp(coef)	exp(-coef)	lower 0.95	upper 0.95
group	0.25	3.999	0.109	0.575
log WBC	5.42	0.184	2.808	10.478

R-square = 0.671 (max possible = 0.988).
Likelihood ratio test = 46.7 on 2 d.f., $p = 7.19e^{-11}$
Wald test = 33.6 on 2 d.f., $p = 5.06e^{-8}$
Score (log rank) test = 46.1 on 2.d.f., $p = 9.92e^{-11}$

Model 3:
$n = 42$

	coef	exp(coef)	se (coef)	Z	p
group	−2.375	0.0935	1.705	−1.393	1.6^{-01}
log WBC	1.555	4.735	0.399	3.900	$9.6e^{-05}$
group: log WBC	0.318	1.374	0.526	0.604	$5.5e^{-01}$

	exp(coef)	exp(-coef)	lower 0.95	upper 0.95
group	0.093	10.750	0.00329	2.63
log WBC	4.735	0.211	2.16741	10.34
group:log WBC	1.374	0.728	0.49017	3.85

R-square = 0.674 (max possible = 0.988)
Likelihood ratio test = 47.1 on 3 d.f., $p = 3.36e^{-10}$
Wald test = 32.4 on 3 d.f. $p = 4.33e^{-7}$
score (log rank) test = 49.9 on 3 d.f., $p = 8.54e^{-11}$

We now discuss the output for the three models shown here. All three models use the same remission times on 42 subjects. the outcome variable for each model is the same time in weeks until a subject goes out of remission. However, the independent variables are different for each model. Model 1 contains only the group variable indicating whether a subject is in treatment or control group. This model can be used in comparison with model 2 to evaluate the potential confounding effect of the variable log WBC. The coefficient for this model is -1.57 indicating that the hazard is decreased (survival is increased) for treatment group. p-value for Z-test is 0.00014 indicating high significance. 95% confidence interval for hazards ratio does not contain 1 showing again significance at 5% level of significance. All the three tests, viz., likelihood ratio, Wald and Score (log rank) indicate high significance as p-values are very small.

We now look at model 2 which contains two variables: the group and log WBC. Our goal is to describe the effect of group adjusted for log WBC. We see that the coefficient for group is -1.39. That is, there is reduction in the hazard (i.e. increase in the survival) when log WBC is introduced in the model. Note that log WBC itself is significant. The chi-square statistic for likelihood ratio test for model 2 is 46.7 with 2 d.f. and that for model 1 is 16.4 with 1 d.f. Thus, the observed difference $46.7 - 16.4 = 30.3$ is realisation of chi-square with 1 d.f. This can be used to test whether it is worthwhile to introduce the log WBC in the model. The high significance of this value indicates that it is so.

Notice that for model 3, the likelihood ratio chi-square statistic is 47.1 with 3 d.f. and for model 2 it is 46.7 with 2 d.f. The difference 0.4 is realisation of chi-square with 1 d.f. It is not significant. Hence it is not worthwhile to introduce interaction term (group $\times$ log WBC) in the model.

The analysis of the output for the three models has led to the conclusion that model 2 is better than the remaining two models.

R-commands for analysis of PH model discussed above are given in the appendix of this chapter.

8.4 Test for Constant of Proportionality in PH Model

We shall describe a commonly used graphical method which gives some feel about the validity of the proportional hazards assumption. If the assumption of proportional hazards is accepted, we may proceed to test the hypothesis that the constant of proportionality (δ) has some specified value and to construct confidence intervals for it.

A. Graphical Test for Proportional Hazards Assumption:

Consider the simplest case of a single covariate at two levels denoted by $z = 0$ and $z = 1$. Let X and Y denote the lifetimes of the subject for the two values of the covariate Z. Let $\overline{F}$ and $\overline{G}$ be the survival functions and h_G and h_F be the hazard functions of X and Y respectively. Under the proportional hazards model we have

$$h_G(t) = \delta h_F(t) \tag{8.4.1}$$

where $\delta > 0$ is the constant of proportionality. Equation (8.4.1), in terms of survival functions is

$$\overline{G}(t) = [\overline{F}(t)]^{\delta}. \tag{8.4.2}$$

Under commonly used log linear link function $\delta = e^{\beta}$ where β is the regression coefficient.

Using (8.4.2) we have

$$-\ell n(-\ell n(\overline{G}(t))) = -\ell n(-\ell n(\overline{F}(t))) - \ell n(\delta) = -\ell n(-\ell n(\overline{F}(t))) - \beta. \tag{8.4.3}$$

Now consider the estimated transformed survival curves corresponding to two groups indexed by 0 and 1.

The difference between their transformed survival curves is β, a constant not depending on t. So if we plot the transformed survival curves for the two groups on the same plot, they should be parallel if the PH assumption holds.

Illustration 8.4

For the following two data sets, use the graphical test to validate the assumption of proportional hazards.

(I) The following is a data set from randomised clinical trial investigating prednisolone therapy reported by Kirk et al. (1980) and discussed in Pocock

(1986). These are survival times in months until death from chronic active hepatitis patients (+ denotes censored data).
Treatment group:
2, 6, 12, 54, 56^+, 68, 89, 96, 96, 125^+, 128^+, 131^+, 140^+, 141^+, 143, 145^+, 146, 148^+, 162^+, 168, 173^+, 181^+.
Control group:
2, 3, 4, 7, 10, 22, 28, 29, 32, 37, 40, 41, 54, 61, 63, 71, 127^+, 140^+, 146, 158^+, 167^+, 182^+.
(II) The following data set is a subset of remission times of leukemia patients studied by Freireich et al. (1963) divided into a treatment group and a control group.
Treatment group: 6, 6, 6, 7, 10, 13, 16, 2, 23, 6^+, 9^+, 10^+, 11^+, 17^+, 19^+, 20^+, 25^+, 32^+, 34^+, 35^+.
Control group: 1, 1, 2, 2, 3, 4, 4, 5, 5, 8, 8, 8, 8, 11, 11, 12, 12, 15, 17, 22, 23.
Data Set 1:

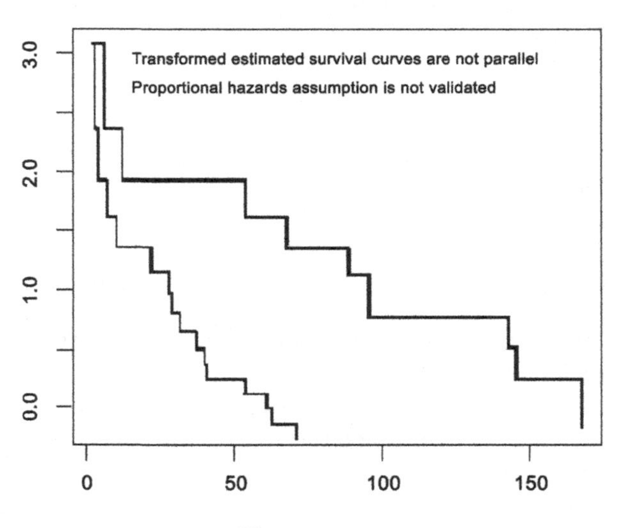

Figure 8.4

Data Set 2:

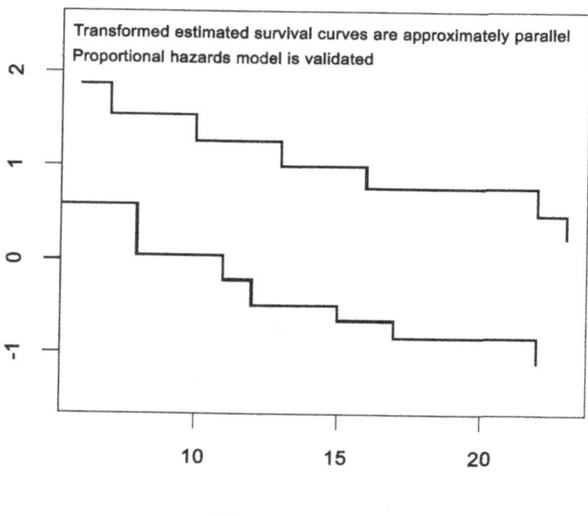

Figure 8.5

R-commands for plotting curves for data set 1 are given in the appendix of this chapter.

Analytical Test for Constant of Proportionality (Deshpande, Frey and Ozturk (2005))

Let us assume that the covariate has only two levels, e.g., presence or absence of a treatment, denoted by $z = 1$ and 0 respectively. Let X and Y denote the lifetimes of a subject for these two values of z. If $\overline{F}$ and $\overline{G}$ are the survival functions of X and Y respectively then, under the proportional hazards model these two functions are related as

$$\overline{G}(t) = (\overline{F}(t))^{\delta}, \quad \delta > 0 \ \forall \ t,$$

since this leads, in case of continuous distribution, to

$$h_G(t) = \delta \ h_F(t),$$

the proportionality of the two hazard rates. The value $\delta = 1$ indicates that X and Y have the same probability distribution. The Wilcoxon-Mann-Whitney or the log rank test may be used to test the null hypothesis that $\delta = 1$. However, more generally the proportional hazards model may be an accepted fact and the null hypothesis $H_0 : \delta = \delta_0$ (δ_0 specified but not necessarily 1) may be required to be tested. Or within this model δ may be unknown and required to be estimated. One can then base both the test of hypotheses and the confidence intervals on the Wilcoxon-Mann-Whitney statistic.

Let $X_1, X_2, \cdots, X_n$ and $Y_1, Y_2, \cdots, Y_m$ be two independent random samples from F and G respectively. At this time we assume that there is no censoring. Define

$$U = \frac{1}{nm} \sum_{i=1}^{n} \sum_{j=1}^{m} \phi(X_i, Y_j)$$

where

$$\phi(X_i, Y_j) = \begin{cases} 1, \text{ if } X_i \leq Y_j \\ 0, \text{ otherwise.} \end{cases}$$

Then

$$E(U) = E\phi(X_i, Y_j) = \int_0^\infty \overline{G}(t) dF(t) = \int_0^\infty (1 - F(t))^\delta dF(t) = \frac{1}{\delta + 1}.$$

Hence under $H_0 : \delta = \delta_0$ we have $E(U) = \frac{1}{\delta_0 + 1}$ and under $H_1 : \delta < \delta_0$ we have $E(U) > \frac{1}{\delta_0 + 1}$. Hence a test which rejects for large values of U is reasonable in this context. Similarly one can find that

$$V(U) = \frac{1}{nm} \frac{\delta}{(\delta+1)^2} + \frac{n-1}{nm} \frac{\delta}{(\delta+1)^2(\delta+2)} + \frac{m-1}{nm} \frac{\delta^2}{(2\delta+1)(\delta+1)}$$

and

$$\lim_{n \to \infty} V(\sqrt{N} U) = \frac{1}{(1-\lambda)} \frac{\delta}{(\delta+1)^2(\delta+2)} + \frac{1}{\lambda} \cdot \frac{\delta^2}{(2\delta+1)(\delta+1)^2}$$

where $N = n + m$ and $\lambda = \lim_{N \to \infty} \frac{n}{N}$.

U being a U-statistic has asymptotic normal distribution. Therefore a test based on the standardised version

$$U^* = \frac{U - E_{H_0}(U)}{\sqrt{V_{H_0}(U)}}$$

may use the critical point from the $N(0, 1)$ distribution.

Furthermore, it can be seen that the expression for $\lambda = \frac{1}{2}$

$$\lim_{N \to \infty} V(\sqrt{N} \;\; U)$$

attains its maximum value, which is $1/3$ at $\delta = 1$. Hence one can construct conservative confidence intervals for δ as

$$\left[\frac{1}{U + z_{\alpha/2}\frac{1}{\sqrt{3N}}} - 1, \frac{1}{U - z_{\alpha/2}\frac{1}{\sqrt{3N}}} - 1\right]$$

by inverting the probability statement regarding U. It is also possible to estimate the asymptotic variance by replacing $\frac{1}{1+\delta}$ by U and $\frac{\delta}{1+\delta}$ by $1 - U$ in its expression. Then one will get approximate confidence intervals

$$\left[\frac{1}{U + z_{\alpha/2}\frac{\hat{\sigma}_U}{\sqrt{3N}}} - 1, \frac{1}{U - z_{\alpha/2}\frac{\hat{\sigma}_U}{\sqrt{3N}}} - 1\right]$$

where $\hat{\sigma}_U^2$ is the estimate of σ_U^2 given by

$$\hat{\sigma}_U^2 = U(1-U)\left[\frac{U}{(1-\lambda)(U+1)} + \frac{1-U}{\lambda(2-U)}\right].$$

Illustration 8.5

The following is a survival data set from 30 patients with AML (Acute Myelogenous Leukemia). The following possible prognostic factors are considered:

$$x_1 = \begin{cases} 1 \text{ if patient } \geq 50 \text{ years old} \\ 0 \text{ otherwise} \end{cases}$$

$$x_2 = \begin{cases} 1 \text{ if celluarity of marrow clot section is } 100\% \\ 0 \text{ otherwise.} \end{cases}$$

Table 8.2

Survival times and data of two possible prognostic factors of 30 AML patients

Survival time	x_1	x_2	Survival time	x_1	x_2
18	0	0	8	1	0
9	0	1	2	1	1
28^+	0	0	26^+	1	0
31	0	1	10	1	1
39^+	0	1	4	1	0
19^+	0	1	3	1	0
45^+	0	1	4	1	0

Table 8.2 (Cont'd)

Survival Time	x_1	x_2	Survival Time	x_1	x_2
6	0	1	18	1	1
8	0	1	8	1	1
15	0	1	3	1	1
23	0	0	14	1	1
28^+	0	0	3	1	0
7	0	1	13	1	1
12	1	0	13	1	1
9	1	0	35^+	1	0

1. Fit a null semi-parametric pH model with log linear link to the data.
2. Fit a semi-parametric pH model with log linear link considering the covariate x_1 only.
3. Fit a semi-parametric pH model with log linear link considering the covariate x_2 only.
4. Fit a semi-parametric pH model with log linear link considering the covariates x_1 and x_2 both.
5. Select the best model from the above models.

Solution:

Output from R

Null Model:

We consider the pH model with no covariate. We get likelihood for this model in the output.

log likelihood $= -65.32214$

$n = 30$

Model 1:

We consider pH model with single covariate x_1. The following is the output:

$n = 30$, number of events $= 23$

	Coef.	exp (coef)	se (coef)	Z	p
x_1	0.9822	2.6702	0.4489	2.188	0.0287*

	exp (coef)	exp (-coef)	lower 0.95	upper 0.95
x_1	2.67	0.3745	1.108	6.436

Concordance = 0.64 (se = 0.061)
R-square = 0.156 (max possible = 0.987)
Likelihood ratio test = 5.08 on 1 d.f., $p = 0.02415$
Wald test = 4.79 on 1 d.f., $p = 0.02867$
Score (logrank) test = 5.13 on 1 d.f., $p = 0.02355$
loglik = $-65.32214, -62.78010$
Interpretation of this output is similar to that of illustration 8.4. In this output we have extracted log likelihood also. There are two entries, first is log likelihood of null model and the second is log likelihood of model 1. The difference $-2^* \log L$ for null model - $(-2^* \log L)$ for model 1 is 5.081 which is realisation of the chi-square statistic with one degree of freedom. This is precisely the value of likelihood ratio statistic in the above output. From the output it is clear that x_1 is significant at 5% level of significance.
Model 2:

We consider pH model with single covariate x_2. The following is the output: $n = 30$, number of events = 23

	Coef.	exp (coef)	se (coef)	Z	p
x_2	0.1884	1.2073	0.4314	0.437	0.662

	exp (coef)	exp (-coef)	lower 0.95	upper 0.95
x_2	1.207	0.8283	0.5183	2.812

Concordance = 0.506 (se = 0.061)
R-square = 0.006 (max possible = 0.987)
Likelihood ratio test = 0.19 on 1 d.f., $p = 0.6605$
Wald test = 0.19 on 1 d.f., $p = 0.6624$
Score (logrank) test = 0.19 on 1 d.f., $p = 0.6619$
loglik = $-65.32214, -65.22568$
The difference $-2^* \log L$ for null model - $(-2^* \log L)$ for model 2 is 0.1929 which is realisation of the chi-square statistic with one degree of freedom.

This is the value of likelihood ratio statistic in the above output.
From the output we see that x_2 is not significant.
Model 3:

We consider the pH model with both the covariates included in the model.
$n = 30$, number of events = 23

	Coef.	exp (coef)	se (coef)	Z	p
x_1	1.0463	2.8472	0.4581	2.284	0.0224*
x_2	0.3586	1.4313	0.4401	0.815	0.4152

	exp (coef)	exp (-coef)	lower 0.95	upper 0.95
x_1	2.847	0.3512	1.1600	6.988
x_2	1.431	0.6987	0.6041	3.391

loglik = $-65.32214, -62.44106$
Concordance = 0.648 (se = 0.068)
R-square = 0.175 (max possible = 0.987)
Likelihood ratio test = 5.76 on 2 d.f., $p = 0.05607$
Wald test = 5.4 on 2 d.f., $p = 0.06718$
Score (logrank) test = 5.77 on 2 d.f., $p = 0.05594$
The difference $-2^* \log L$ for null model - $(-2^* \log L)$ for model 3 is 5.76216 which is realisation of the chi-square statistic with two degree of freedom. This is the value of likelihood ratio statistic in the above output.

The best model is model 1.

For R-commands see the appendix of this chapter.

8.5 Frailty Model

The proportional hazards model of Cox has been developed to incorporate covariates with fixed known levels in a simple way. However, in many situations the levels (values) of the covariates may not be fixed and not known either. They may then be regarded as unknown realisations of random variables with known probability distributions. In classical linear regression theory also random effects, along with fixed effects are accommodated in mixed effect models. In survival analysis the random part of the regression is called 'frailty'. This situation arises when the experimental units (patients, organs, or other entities) are not and cannot be perfectly matched. They are subject to heterogeneity which cannot even be measured. This may arise because of different genetic makeup, or differences in their frailty or risk processes due to unidentifiable environmental factors. Sometimes we may be able to group the patients according to perceived homogeneity (which may exist among siblings, or patients living in close proximity, etc.).

The fundamental concern is that units which are subjected to treatments may not be homogeneous, and this lack of homogeneity may not be observable.

Univariate Frailty Models

The standard multiplication model which incorporates frailty is given in terms the hazard (or the failure) rate as

$$\lambda(t|Z, \underline{x}) = Z\lambda(t|\underline{x}), \tag{8.5.1}$$

where Z is a random variable providing the values of frailty from individuals and $\underline{x}$ is the vector of value of the usual regressor variable which may affect the hazard rate. Further, if we assume the Cox proportional hazards model then

$$\lambda(t|Z, \underline{x}) = Z \cdot \lambda_0(t) \cdot \exp\{\underline{\beta}'\underline{x}\} \tag{8.5.2}$$

where $\lambda_0(t)$ is the baseline hazard rate and the exponential factor is the link function which connects the values of the regressor variables through the regression coefficients to the overall hazard rate. The point to note is that Z is a random variable that makes a patient more prone or more vulnerable when $Z > 1$ and less prone when $Z < 1$.

Gamma Frailty Model

Many distributions have been proposed for the frailty random variable Z, the gamma distribution being the most common.

Let us suppose that T_{ij} are the lifetimes of the j-th individual in the i-th frailty group, $j = 1, 2, \cdots, k_i$; $k_1, k_2 \cdots, k_n$ being the number of members in the n groups.

Following the expression (8.5.2) one can write the joint survival function of all the individuals with the i-th shared frailty as

$$\begin{aligned}\overline{F}(t_{i_1},\cdots,t_{i_{k_i}}) &= P(T_{i_1} > t_{i_1},\cdots,T_{i_{k_i}} > t_{i_{k_i}}) \\ &= \int_0^\infty \prod_{j=1}^{k_i} P(T_{ij} > t_{ij}|z_i)g(z_i)dz_i\end{aligned}$$

where g is the common p.d.f. of the frailty variable Z_i's. We assume the gamma distribution with parameters $(\theta, 1/\theta)$, to avoid identifiability problem, i.e.

$$g(z) = z^{(\frac{1}{\theta}-1)}e^{-\frac{z}{\theta}}/\Gamma(\frac{1}{\theta})\cdot\theta^{1/\theta},\quad \theta > 0.$$

This distribution has $E(Z) = 1$ and $V(Z) = \theta$. Then

$$\overline{F}(t_{i_1},\cdots,t_{i_{k_i}}) = \left(1+\theta\sum_{j=1}^{k_i}\Lambda_0(t_{ij})\exp\{\underline{\beta}'\underline{x}_{ij}\}\right)^{-\theta} \tag{8.5.3}$$

where $\Lambda_0(t) = \int_0^t \lambda_0(u)du$, the cumulative failure rate of the baseline hazard rate. Since the individuals in different groups, with different frailty values, are regarded as independent, the complete survival probability is written as the product of the expression in (8.5.3) over $i = 1,\cdots,n$.

We carry out maximum likelihood estimation of the parameters in this model.

Define $\delta_{ij} = 1$ if the j-th individual in the i-th groups is a complete life and $\delta_{ij} = 0$, if it is censored. Also $D_i = \sum_{j=1}^{k_i}\delta_{ij}$. The data actually consists of $(t_{ij},\delta_{ij},X_{ij}), i = 1,\cdots,n, j = 1,\cdots,k_i$. The log likelihood is then

$$\begin{aligned}L(\theta,\underline{\beta}) = \sum_{i=1}^{n} D_i\log\theta &- \Gamma(\tfrac{1}{\theta}) + \Gamma(\tfrac{1}{\theta}+D_i) \\ &-(\tfrac{1}{\theta}+D_i)\log\left[1+\theta\sum_{i=1}^{k_i}\Lambda_0(t_{ij})\exp(\underline{\beta}'\underline{x}_{ij})\right] \\ &+\sum_{j=1}^{k_i}\delta_{ij}\{\underline{\beta}'\underline{x}_{ij}+\log(\lambda_0(t_{ij}))\}.\end{aligned} \tag{8.5.4}$$

If one further assumes a parametric form for the baseline hazard rate λ_0 (say, exponential, Weibull, etc.) then direct numerical maximisation will give the maximum likelihood estimators. In case a parametric form is not assumed for λ_0 then the E-M algorithm helps to obtain the maximum likelihood estimators. For details see Klein and Moeschberger (2003) or Hanagal (2011).

Positive Stable Frailty Distribution

A stable distribution may have a scale parameter $\delta > 0$ and an index parameter $\alpha < 1$. We again impose the condition $\delta = \alpha$ for the sake of identifiability. The probability density function is quite complicated, but its Laplace transform is $L(s) = \exp(-s^\alpha)$. Then

$$P(T_{ij} > t) = \exp\{-(\Lambda_0(t_{ij}))^\alpha \exp(\alpha \cdot \underline{\beta}' x_{ij})\}.$$

If we write the p-th derivative of the Laplace transform as a polynomial

$$L^{(1)}(p) = (-1)^p \exp(-s^\alpha) \sum_{m=1}^{p} c_{p,m} \theta^m s^{m\alpha - p}$$

where $c_{p,m}$ themselves are polynomials in α of degree m. These are defined recursively by

$$c_{p,p} = 1, \quad c_{p,1} = \frac{\Gamma(p-\alpha)}{\Gamma(1-\alpha)}$$

and $c_{pm} = c_{p-1,m-1} + c_{p-1,m}((p-1) - m\alpha)$. Using these one can write the complete likelihood function as

$$\prod_{i=1}^{n} \prod_{j=1}^{k_i} (\lambda_0(t_{ij}))^{\delta_{ij}} \exp\{\delta_{ij} \underline{\beta}' \underline{x}_{ij}\} \ \exp\left\{ - \sum_{j=1}^{k_i} \Lambda_0(t_{ij}) \exp(\underline{\beta}' x_{ij}) \right\}^\alpha$$

$$\times \sum_{m=1}^{D_i} c_{D_i,m} \alpha^m \left[\sum_{j=1}^{k_i} \Lambda_0(t_{ij}) \exp \underline{\beta}' y_{ij} \right]^{m\alpha - D_i} . \qquad (8.5.5)$$

This can then be maximised by using the E-M algorithm.

Illustration 8.6

Consider the well-known kidney data set which contains the recurrence times to kidney infection for 38 patients using portable dialysis equipment. The dataset is available in R-package "parfm". We use the subset of this data set, say mydata, obtained by deleting column number 7 of the data set kidney. The first 6 observations of the data set mydata look like the following:

id	time	status	age	sex	disease
1	8	1	28	1	Other
1	16	1	28	1	Other
2	23	1	48	2	GN
2	13	0	48	2	GN
3	22	1	32	1	Other
3	28	1	32	1	Other

Each observation corresponds to a kidney; the variable id is the patient's code. The time from insertion of catheter to infection or censoring is stored in "time" while "status" is 1 when infection has occurred and 0 for censored observation (catheters may be removed for reasons other than infection). Three covariates are available: age, age of patient in years, sex (1 for males and 2 for females) and disease type (GN, AN, PKD and other).

1. Fit a model with the positive stable frailty distribution ("possta") and exponential baseline hazard function. Interpret the output of the model.
2. Fit a model with the inverse Gaussian distributed frailties ("ingau") and "weibull" baseline hazard function. Interpret the output of the model.
3. Compare the two models in parts 1 and 2 above on the basis of AIC and BIC criteria.

Solution:

1. R-output of the model:

parfm(Surv(time, status) sex+ age, cluster="id", data=mydata, dist= "exponential", frailty= "possta");

is:

Frailty distribution: positive stable

Baseline hazard distribution: Exponential

Log likelihood: −337.132

ESTIMATE SE p-val

nu 0.000

lambda 0.012

sex −0.885

age 0.004

Signif. codes: 0 '***' 0.001 '**' 0.01 '*' 0.05 '.' 0.1 ' ' 1

Kendall's Tau: 0

For the frailty distribution under consideration, both the mean and variance are undefined. Therefore the heterogeneity parameter, nu, in the above output does not correspond to the variance of the frailty term. Since the value of heterogeneity parameter for the problem under consideration is zero, the model with positive frailty distribution converges to a solution which is not valid.
The problem can be fixed by changing the optimisation method. By default it is set to "BFGS". Change this to "Nelder-Mead" by adding the argument method = "Nelder-Mead" to the model.
The modified model is:
parfm(Surv(time, status) ~ sex+ age, cluster="id", data=mydata, dist= "exponential", frailty= "possta", method ="Nelder-Mead");
R-output of this model is:
Frailty distribution: positive stable
Baseline hazard distribution: Exponential
Log likelihood: −336.182

	ESTIMATE	SE	p-val
nu	0.112	0.084	
lambda	0.014	0.008	
sex	−0.951	0.348	0.006 **
age	0.004	0.011	0.694

Signif. codes: 0 '***' 0.001 '**' 0.01 '*' 0.05 '.' 0.1 ' ' 1
Kendall's Tau: 0.112
Interpretation of the output:
The estimated value of the heterogeneity parameter (nu) is 0.112 and its standard error (SE) is 0.084. lambda is baseline hazard. Its estimated value is 0.014 and its SE is 0.008. The covariate sex is highly significant with a p-value of 0.006. Conditional on the frailty of patient and on the age the hazard of infection for a female at any time t is estimated to be exp (−0.951). The covariate age is not significant.
2. R-output of the model:
parfm(Surv(time, status) sex+ age, cluster="id", data=mydata, dist= "weibull", frailty= "ingau")
is: Frailty distribution: inverse Gaussian
Baseline hazard distribution: Weibull
Log likelihood: −333.314

ESTIMATE	SE	p-val		
theta	0.677	0.540		
rho	1.145	0.148		
lambda	0.013	0.011		
sex	−1.481	0.432	0.001	***
age	0.006	0.012	0.655	

Signif. codes: 0 '***' 0.001 '**' 0.01 '*' 0.05 '.' 0.1 ' ' 1
Kendall's Tau: 0.181

The conclusions drawn from this model are essentially the same as that of modified model in part 1.
3. R-output of the model selection command:
mydata.select<-select.parfm(Surv(time,status) ~ sex + age, cluster = "id", data = mydata, dist = c("exponential", "weibull"), frailty=c("ingau", "possta"))
is:
AIC:

	ingau	possta
exponential	675.699	682.264
weibull	676.627	682.315

BIC:

	ingau	possta
exponential	685.022	691.587
weibull	688.281	693.969

In this particular example the exponential baseline seems to be a good choice.
For R-commands see the appendix of this chapter.
Illustration 8.7

Fit the semi-parametric model with gamma frailties via the coxph() function using the data set "kidney" and write your conclusions.
Solution:
R-output of semi-parametric model with gamma frailties:

	coef	se(coef)	se2	Chisq	DF	p
sex	−1.58323	0.4594	0.3515	11.88	1.0	0.00057
age	0.00522	0.0119	0.0088	0.19	1.0	0.66000

frailty(id, distrubution) 22.96 12.9 0.04100
Variance of random effect = 0.408 I-likelihood = −181.6
Degrees of freedom for terms = 0.6 0.6 12.9
Likelihood ratio test = 46.5 on 14.1 df, p = 2.37e-05 n = 76

Conclusions: Estimates of covariates sex and age are quite similar to the estimates obtained from modified frailty model in part 1, and also in the frailty model in part 2 of Illustration 8.6. However, the frailty variances are different because of the difference in how the baseline hazard is treated.

8.6 Accelerated Lifetime (ALT) Model

The more reliable a device is, the more difficult to measure its reliability. This is because the reliable devices last a long time under actual operating conditions. In some situations many years will be required to obtain data on failure times under actual field conditions. A solution to this problem is accelerated testing. For testing purposes the devices are made to operate under harsher conditions which will hasten their failure. The resulting decrease in the lifetimes lead to experiments requiring shorter times and hopefully less cost.

The main purpose of accelerated life testing (ALT) is to make inference about the life distribution and its parameters or quantities from the data obtained through accelerated testing or increased stress conditions. We model the relationship between the life distribution at actual use conditions and at accelerated life conditions through a regression like link.

The ALT Model

It is possible to bring in the effect of the harsher conditions through the proportional hazards model. However, the ALT model envisages the effect of regressor variables or covariates specifying the ALT conditions through the scale parameter.

Let $S_0(t)$ be the baseline survival function (perhaps for operation at actual field conditions). The vector $\underset{\sim}{z}$ brings in the changes for the ALT and $\underset{\sim}{\beta}$ be the regression coefficients which link the changes to the baseline.

Then

$$S(t/\underline{\beta}, \underline{z}) = S_0(t/\exp(\underline{t}'\underline{z}))$$

will be the survival function at the ALT conditions. Note that $\underline{z} = \underline{0}$ takes us back to the baseline survival function. One can also write the random variable representing the lifetime at the conditions $\underline{z}$ as

$$T(\underline{z}) = e^{\underline{\beta}'\underline{z}} \cdot T_0$$

where T_0 is the lifetime at baseline conditions. Equivalently, we have a linear model in $\log T(\underline{z})$ as

$$\log T(\underline{z}) = \log T_0 + \beta_1 z_1 + \cdots \beta_p z_p$$

since we are dealing with positive valued random variables.

Also, if the hazard rate under standard baseline condition is $h_0(t)$ then

$$h(t/\underline{\beta}, \underline{z}) = e^{-\underline{\beta}'\underline{z}} h_0(t/\exp(\underline{\beta}'\underline{z}))$$

at the ALT conditions.

The Weibull Distribution

It may be noted that if the probability distribution belongs to the Weibull model then the hazard rate under the ALT model is

$$\begin{aligned} h(t|\underline{\beta}, \underline{z}) &= e^{-\underline{\beta}'\underline{z}} \lambda \cdot \gamma \left(\frac{t}{\underline{\beta}'\underline{z}} \right)^{\gamma-1} \\ &= (e^{\underline{\beta}\underline{z}})^{\gamma} \lambda \cdot \gamma . t^{\gamma-1} \end{aligned}$$

which is proportional to $h_0(z) = \lambda \cdot \gamma t^{\gamma-1}$. Hence in the Weibull case (and only in this case) the proportional hazards model and the ALT model are equivalent.

The Log Linear ALT Model

Define

$$Y = \log T(\underline{z}) = \mu + \underline{\beta}'\underline{z} + \sigma X$$

where the $\underline{\beta}$ is the vector of regression coefficients, X the error, μ and σ the location and scale parameters which connect the error to $Y = \log T(\underline{z})$. The scale model given in (2) above is related to this model if we let S_0 be the survival function of $\exp(\mu + \sigma X)$. Many models have been suggested for the distribution of X or $\exp(\mu + \sigma X)$. The Weibull or the log logistic model are two such models. In these and in other suitable models, the parameters are estimated by the maximum likelihood methods. Besides the estimated asymptotic covariance matrix of the estimators will be required for standard errors and for obtaining confidence intervals. All of these need to be obtained by appropriate numerical procedures such as the Newton-Raphson method.

Illustration 8.8:

We consider here a parametric model with covariates and failure times following the survival function of the form described in part 4 above. For the analysis we use the data, "Death Times of Male Laryngeal Cancer Patients" ("larynx"). Kardaun (1983) reports data on 90 mails diagnosed with cancer of the larynx during the period 1970-1978 at the Dutch hospital. Times recorded are the intervals (in years) between first treatment and either death or end of the study (January 1, 1983). Also recorded are the patient's ages at the time of diagnosis, the year of diagnosis and the stage of the patient's cancer. The four stages of disease are: 1=stage 1, 2=stage 2, 3=stage 3, 4=stage 4. The stages are ordered from least serious to most serious. The data are available in the R-package KMsurv.

Fit accelerated model given by:

$$Y = ln(X) = \mu + \gamma_1 Z_1 + \gamma_2 Z_2 + \gamma_3 Z_3 + \gamma_4 Z_4 + \sigma W$$

where $Z_1, \cdots, Z_3$ are the indicators of stage 2, stage 3 and stage 4 disease, respectively and Z_4 is the age of the patient.

The main choice to be made is which distribution to use.

(1) Assume Weibull distribution and fit the model. Prepare a table showing estimates of the parameters: Intercept, Log Scale and Z_1, Z_2, Z_3, Z_4 with the corresponding std. errors and p-values. Interpret the results. Obtain AIC of the model.

(2) Assume Exponential distribution and fit the model. Prepare a table showing estimates of the parameters: Intercept, and Z_1, Z_2, Z_3, Z_4 with the corresponding std. errors and p-values. Obtain AIC of the model.

Compare the two models. Which model is more appropriate?

Solution: (1)

Variable	Parameter estimate	Standard errors	Wald statistics	P-values
Intercept (mu)	3.5288	0.9041		
Log(Scale)	−0.1223	0.1225		
Z1:Stage 2	−0.1477	0.4076	−0.362	0.717
Z2:Stage 3	−0.5866	0.3199	−1.833	0.0668
Z3: Stage 4	−1.5441	0.3633	−4.251	<0.0001
Z4: Age	−0.0175	0.0128	−1.367	0.172

P-values are used to test the hypothesis: $\gamma_i = 0$.

Here we see that the patients with stage 4 disease do significantly worse than the patients with stage 1 disease. It is important to note that as opposed to Cox model where a positive value of the risk coefficient reflects poor survival, here, a negative value of the coefficient is indicative of decreased survival.

The effect of age is not significant.

AIC of the model is 286.8.

(2) For exponential distribution scale is fixed at 1.

Variable	Parameter estimate	Standard errors	Wald statistics	P-values
Intercept(mu)	3.7550	0.9902		
Z1:Stage 2	−0.1456	0.4602	−0.316	0.752
Z2:Stage 3	−0.6483	0.3552	−1.825	0.068
Z3: Stage 4	−1.6350	0.3985	−4.103	<0.0001
Z4: Age	−0.0197	0.0142	−1.388	0.165

The AIC of the exponential distribution is 285.8.

AIC for exponential is lower than AIC for Weibull;

Exponential distribution is more appropriate.

Exponential model is a special case of Weibull. In part 1, the p-value for Log (scale) is 0.318, which is not significant. This also suggests that exponential model is more appropriate.

For R-commands see the appendix of this chapter.

Exercises

Exercise 8.1: Carry out the exact test to determine if the wattage has any influence on the survival distribution of the bulbs using the data of illustration 1.

Exercise 8.2: In a clinical trial described by Kirk et al. (1980) 44 patients with chronic active hepatitis were randomised to the drug prednisolone, or an untreated control group. The survival time of the patients, in months, following enrollment to the trial, was the response variable of interest. The data set, which was given in Pocock (1983), is shown in the following table:

Table 8.3

Survival times of patients suffering from chronic active hepatitis

Prednisolone		Control	
3	131+	2	41
6	140+	3	54
12	141+	4	61
54	143	7	63
56+	145+	10	71
68	146	22	127+
89	148+	28	140+
96	162+	29	146+
96	168	32	158+
123+	173+	37	167+
128+	181+	40	182+

(1) Prepare a plot of survival curves.
(2) Fit semi-parametric proportional hazards model with log linear link function to the data. Interpret the output.
(3) Carry out log rank test of Peto and Peto.
(4) Write your conclusions on the basis of above analysis.

Exercise 8.3: Consider the data set veteran of Exercise 7.3. The variables in the data set are listed as follows:

trt	:	1 = standard, 2 = test
celltype	:	1 = squamous, 2 = smallcell, 3 = adeno, 4 = large
time	:	survival time
status	:	censoring status
karno	:	Karnofsky performance score (0 for worst... 100=good)
diagtime	:	months from diagnosis to randomisation
age	:	in years
prior	:	prior therapy 0 = no, 1 = yes

To this data set semi-parametric proportional hazards model with log linear link function is fitted and relevant information from the output is given below:

Table 8.4

Variable	Coef	Se (coef)	p	Exp (coef)	Lower 0.05	Upper 0.05
Treat	0.290	0.207	0.162	1.336	0.890	2.006
Adino cell	0.789	0.303	0.009	2.200	1.210	3.982
Small cell	0.457	0.266	0.08	1.579	0.937	2.661
Squamous cell	−0.400	0.283	0.157	0.671	0.385	1.167
karno	−0.033	0.006	0.000	0.968	0.958	0.978
diagtime	0.000	0.009	0.992	1.000	0.982	1.018
age	−0.009	0.009	0.358	0.991	0.974	1.010
prior	0.007	0.023	0.755	1.007	0.962	1.054

Use the above information to answer the following:

1. What is the hazard ratio that compares persons with adino type cells with persons with large type cells?
2. What is the hazard ratio that compares persons with adino type cells with persons with squamous type cell?
3. Is there any effect of treatment on survival time?
4. Among the variables karno, diagtime, age, and prior which variables are significant?

Justify your answers.

Exercise 8.4: In the data set on the survival times of multiple myeloma patients there are the following seven explanatory variables:

Age	:	Age of the patient,
Sex	:	Sex of the patient (0 = male, 1 = female),
Bun	:	Blood urea nitrogen,
Ca	:	Serum calcium,
Hb	:	Serum hemoglobin,
Pcells	:	Percentage of placenta cells,
Protein	:	Bence-Jones Protein (0 = absent, 1 = present).

Primary interest is to determine the most appropriate subset of these variables. The first step in this analysis is to fit the null model and semi-parametric proportional hazards models with log linear link that contains one of the seven explanatory variables. A summary of the values of $-2\log L$ for these models are given in the following table.

Table 8.5

Variables in model	$-2\log L$
None	215.940
Age	215.817
Sex	215.906
Bun	207.453
Ca	216.494
Hb	211.068
Pcell	215.875
Protein	213.890

Use the above table to obtain top two most explanatory variables. Justify your answer.

Exercise 8.5: The following data set is a sample from the 1987-1990 Evans County study. Survival times (in years) are given for two study groups, each with 25 participants. Group 1 has no history of chronic disease (CHR=0), and group 2 has a positive history of chronic disease (CHR=1);

Group 1:

12.3+,5.4,8.2,12.2+,11.7,10.0,5.7,9.8,2.6,11.0,9.3,12.1+,6.6,2.2,1.8,10.2, 10.7,11.1,5.3,3.5,9.2,2.5,8.7,3.8,3.0.

Group 2:

5.8,2.9,8.4,8.3,9.1,4.2,4.1,1.8,3.1,11.4,2.4,1.4,5.9,1.6,2.8,4.9,3.5,6.5,9.9,3.6,5.2, 8.8,7.8,4.7,3.9

Carry out graphical test for semi-parametric proportional hazards assumption in proportional hazards model with log-linear link.

Exercise 8.6: Twelve patients were recruited in a study of the treatment of cirrhosis of the liver. The patients were randomised to receive either a placebo or a new treatment that was referred as Liverol. Six patients were allocated to placebo and six to Liverol. Age and baseline value of the patient's bilirubin level were recorded at the entry. The natural log of the bilirubin value will be used in the analysis. The variables measured are summarised below:

Time : Survival time of patients in days,
Status : Event indicator (0 = censored, 1 = uncensored),
Treat : Treatment group (0 = Placebo, 1 = Liverol),
Age : Age of the patient in years,
Lbr : bilurubin level.

The following table shows the data set:

Table 8.6

Patient	1	2	3	4	5	6	7	8	9	10	11	12
Time	281	604	457	384	341	842	1514	162	1121	1411	814	1071
Status	1	0	1	1	0	1	1	0	1	0	1	1
Treat	0	0	0	0	0	0	1	1	1	1	1	1
Age	46	57	56	65	73	64	69	62	71	69	77	58
Lbr	3.2	3.1	2.2	3.9	2.8	2.4	2.4	2.4	2.5	2.3	3.8	3.1

Fit a semi-parametric proportional hazards model with log-linear link with

(1) three covariates Treat, Age and natural log of lbr.
(2) two covariates Treat, and natural log of lbr.

Write your conclusions on the basis of fitted two models.

Exercise 8.7: Using the data named "mydata" of Illustration 8.6 to do the following:

1. Fit a model with the frailty distribution gamma and exponential baseline hazard function. Interpret the output of the model.
2. Fit a model with the "loglogistic" frailty distribution and "weibull" baseline hazard function. Interpret the output of the model.

Exercise 8.8: ALT model: Log logistic distribution is an alternative to the "Weibull" distribution. Use the data "larynx" of Illustration 8.8. Fit accelerated model given by:

$$Y = \ln(X) = \mu + \gamma_1 Z_1 + \gamma_2 Z_2 + \gamma_3 Z_3 + \gamma_4 Z_4 + \sigma W$$

where $Z_1, \cdots, Z_3$ are the indicators of stage 2, stage 3 and stage 4 disease, respectively and Z_4 is the age of the patient. Employ the Log logistic model ("loglogistic" is R-notation) to study the effects of stage and age. Prepare a table showing estimates of the parameters: Intercept, Log Scale and Z_1, Z_2, Z_3, Z_4 with the corresponding std. errors and p-values. Interpret the results. Obtain AIC of the model. Compare Log logistic and Weibull models on basis of AIC and p-values of the variables age, Z_1, Z_2, Z_3, Z_4.

Appendix

Cox's Proportional hazards model: Leukemia remission data

```
>  library(survival);
> time<-c(6,6,6,7,10,13,16,22,23,6,9,10,11,17,19,20,25,32,32,
34, 35, 1,1,2,2,3,4,4,5,5,8,8,8,8,11,11,12,12,15,17,22,23);
> length(time);
[1] 42
> status<-c(rep(1,9),rep(0,12),rep(1,21));
> group<-c(rep(1,21),rep(0,21));
> logWBC<-
 c(2.31,4.06,3.28,4.43,2.96,2.88,3.60,2.32,2.57,3.20,2.80,2.70,
 2.60,2.16, 2.05, 2.01,1.78,2.20,2.53,1.47,1.45,2.80,5.00,4.91,
 4.48,4.01,4.36,2.42,3.49,3.97,3.52,3.05,2.32,3.26,3.49,2.12,
 1.50,3.06,2.30,2.95,2.73,1.97);
> length(logWBC);
[1] 42
> leukemia<-data.frame(time,status,group,logWBC);

> leukemia1<-transform(leukemia,int=group*logWBC);
> leukemia1;

   time status group logWBC  int
1     6      1     1   2.31 2.31
2     6      1     1   4.06 4.06
3     6      1     1   3.28 3.28
```

4	7	1	1	4.43	4.43
5	10	1	1	2.96	2.96
6	13	1	1	2.88	2.88
7	16	1	1	3.60	3.60
8	22	1	1	2.32	2.32
9	23	1	1	2.57	2.57
10	6	0	1	3.20	3.20
11	9	0	1	2.80	2.80
12	10	0	1	2.70	2.70
13	11	0	1	2.60	2.60
14	17	0	1	2.16	2.16
15	19	0	1	2.05	2.05
16	20	0	1	2.01	2.01
17	25	0	1	1.78	1.78
18	32	0	1	2.20	2.20
19	32	0	1	2.53	2.53
20	34	0	1	1.47	1.47
21	35	0	1	1.45	1.45
22	1	1	0	2.80	0.00
23	1	1	0	5.00	0.00
24	2	1	0	4.91	0.00
25	2	1	0	4.48	0.00
26	3	1	0	4.01	0.00
27	4	1	0	4.36	0.00
28	4	1	0	2.42	0.00
29	5	1	0	3.49	0.00
30	5	1	0	3.97	0.00
31	8	1	0	3.52	0.00
32	8	1	0	3.05	0.00
33	8	1	0	2.32	0.00
34	8	1	0	3.26	0.00
35	11	1	0	3.49	0.00
36	11	1	0	2.12	0.00
37	12	1	0	1.50	0.00
38	12	1	0	3.06	0.00
39	15	1	0	2.30	0.00
40	17	1	0	2.95	0.00
41	22	1	0	2.73	0.00
42	23	1	0	1.97	0.00

```
> summary(coxph(Surv(time,status==1) ~ group));
Call:
coxph(formula = Surv(time, status == 1) ~ group)
  n= 42, number of events= 30

         coef exp(coef) se(coef)      z Pr(>|z|)
group -1.5721    0.2076   0.4124 -3.812 0.000138 ***
---
Signif. codes: 0 *** 0.001 ** 0.01 * 0.05 . 0.1   1

      exp(coef) exp(-coef) lower .95 upper .95
group    0.2076      4.817   0.09251    0.4659

Concordance= 0.69  (se = 0.053 )
Rsquare= 0.322   (max possible= 0.988 )
Likelihood ratio test= 16.35  on 1 df,   p=5.261e-05
Wald test            = 14.53  on 1 df,   p=0.0001378
Score (logrank) test = 17.25  on 1 df,   p=3.283e-05

> summary (coxph(Surv(time, status == 1) ~ group + logWBC));
Call:
coxph(formula = Surv(time, status == 1) ~ group + logWBC)
  n= 42, number of events= 30

          coef exp(coef) se(coef)      z Pr(>|z|)
group  -1.3861    0.2501   0.4248 -3.263   0.0011 **
logWBC  1.6909    5.4243   0.3359  5.034  4.8e-07 ***
---
Signif. codes: 0 *** 0.001 ** 0.01 * 0.05 . 0.1   1

       exp(coef) exp(-coef) lower .95 upper .95
group     0.2501     3.9991    0.1088    0.5749
logWBC    5.4243     0.1844    2.8082   10.4776

Concordance= 0.852  (se = 0.062 )
Rsquare= 0.671   (max possible= 0.988 )
Likelihood ratio test= 46.71  on 2 df,   p=7.187e-11
Wald test            = 33.6   on 2 df,   p=5.061e-08
Score (logrank) test = 46.07  on 2 df,   p=9.921e-11
```

```
> summary(coxph(Surv(time,status==1) ~ group + logWBC +
  group*logWBC));
Call:
coxph(formula=Surv(time,status== 1)~group+logWBC+group*logWBC)

  n= 42, number of events= 30

                 coef exp(coef) se(coef)      z Pr(>|z|)
group        -2.37491   0.09302  1.70547 -1.393    0.164
logWBC        1.55489   4.73459  0.39866  3.900 9.61e-05 ***
group:logWBC  0.31752   1.37372  0.52579  0.604    0.546
---
Signif. codes: 0 *** 0.001 ** 0.01 * 0.05 . 0.1  1

             exp(coef) exp(-coef) lower .95 upper .95
group          0.09302    10.7500  0.003288     2.632
logWBC         4.73459     0.2112  2.167413    10.342
group:logWBC   1.37372     0.7280  0.490169     3.850

Concordance= 0.851  (se = 0.062 )
Rsquare= 0.674   (max possible= 0.988 )
Likelihood ratio test= 47.07  on 3 df,   p=3.356e-10
Wald test            = 32.39  on 3 df,   p=4.326e-07
Score (logrank) test = 49.86  on 3 df,   p=8.539e-11
```

Graphical test for proportional hazards model

Data set 1

```
> time<-c(2,6,12,54,56,68,89,96,96,125,128,131,140,141,143,
  145,146,148,162, 168,173,181);
> length(time);
[1] 22
> status<-c(rep(1,4),0,rep(1,4),rep(0,5),1,0,1,0,0,1,0,0);
> length(status);
[1] 22
> treat<-data.frame(time,status);
> library(survival);
```

```
> attach(treat);

The following objects are masked _by_ .GlobalEnv:
    status, time

> surve.treat<-survfit(Surv(time,status)~1);
> summary(surve.treat);
Call: survfit(formula = Surv(time, status) ~ 1)

time n.risk n.event survival std.err lower 95% CI upper 95%CI
   2     22       1    0.955  0.0444        0.871       1.000
   6     21       1    0.909  0.0613        0.797       1.000
  12     20       1    0.864  0.0732        0.732       1.000
  54     19       1    0.818  0.0822        0.672       0.996
  68     17       1    0.770  0.0904        0.612       0.969
  89     16       1    0.722  0.0967        0.555       0.939
  96     15       2    0.626  0.1051        0.450       0.870
 143      8       1    0.547  0.1175        0.359       0.834
 146      6       1    0.456  0.1285        0.263       0.793
 168      3       1    0.304  0.1509        0.115       0.804
> time<-c(2,3,4,7,10,22,28,29,32,37,40,41,54,61,63,71,
  127,140,146,158,167,182);
> status<-c(rep(1,16),rep(0,6));
> surv.ctrl<-survfit(Surv(time,status)~1);
> summary(surv.ctrl);
Call: survfit(formula = Surv(time, status) ~ 1)

time n.risk n.event survival std.err lower 95% CI upper 95%CI
   2     22       1    0.955  0.0444        0.871       1.000
   3     21       1    0.909  0.0613        0.797       1.000
   4     20       1    0.864  0.0732        0.732       1.000
   7     19       1    0.818  0.0822        0.672       0.996
  10     18       1    0.773  0.0893        0.616       0.969
  22     17       1    0.727  0.0950        0.563       0.939
  28     16       1    0.682  0.0993        0.513       0.907
  29     15       1    0.636  0.1026        0.464       0.873
  32     14       1    0.591  0.1048        0.417       0.837
  37     13       1    0.545  0.1062        0.372       0.799
  40     12       1    0.500  0.1066        0.329       0.759
```

```
  41     11      1    0.455  0.1062        0.288       0.718
  54     10      1    0.409  0.1048        0.248       0.676
  61      9      1    0.364  0.1026        0.209       0.632
  63      8      1    0.318  0.0993        0.173       0.587
  71      7      1    0.273  0.0950        0.138       0.540
> time<-c(2,6,12,54,68,89,96,143,146,168);
> survival<-c(0.955,0.909,0.864,0.818,0.770,0.722, 0.626,
  0.547,0.456,0.304);
> modisurv1<-(-log(survival,base=exp(1)));
> modisurv2<-(-log(modisurv1,base=exp(1)));
> plot(time,modisurv2,"s",lwd=2);
> time<-c(2,3,4,7,10,22,28,29,32,37,40,41,54,61,63,71);
> survival<-c(0.955,0.909,0.864,0.818,0.773,0.727,0.682,0.636,
  0.591, 0.545, 0.50,0.455, 0.409,0.364,0.318,0.273);
> modisurv1<-(-log(survival,base=exp(1)));
> modisurv2<-(-log(modisurv1,base=exp(1)));
> points(time,modisurv2,"s",lwd=2);
> text(locator(1),
 "Transformed estimated survival curves are not parallel", cex=0.8);
> text(locator(1),
 "Proportional hazards assumption is not validated",cex=0.8);
```

Illustration 8.6: Cox proportional hazards model: AML data

```
> time<-c(18, 9, 28, 31, 39, 19, 45, 6, 8, 15, 23, 28, 7,
  12, 9, 8, 2, 26, 10, 4, 3, 4, 18, 8, 3,14, 3, 13, 13,35);
> length(time);
[1] 30
> status<-c(1,1,0,1,rep(0,3),rep(1,4),0,rep(1,5),0,rep(1,11),0);
> length(status);
[1] 30
> x1<-c(rep(0,13),rep(1,17));
> length(x1);
[1] 30
> x2<-c(0,1,0,1,1,1,1,1,1,1,0,0,1,0,0,0,1,0,1,0,0,0,1,
  1,1,1,0,1,1,0);
> length(x2);
[1] 30
> library(survival);
```

```
> nullmod<- (coxph(Surv(time,status==1)~1));
> summary(nullmod);
Call:  coxph(formula = Surv(time, status == 1) ~ 1)

Null model
  log likelihood= -65.32214
  n= 30
> fit1<-coxph(Surv(time,status==1)~x1);
> summary(fit1);
Call:
coxph(formula = Surv(time, status == 1) ~ x1)

  n= 30, number of events= 23

     coef exp(coef) se(coef)     z Pr(>|z|)
x1 0.9822    2.6702   0.4489 2.188   0.0287 *
---
Signif. codes: 0 *** 0.001 ** 0.01 * 0.05 . 0.1  1

   exp(coef) exp(-coef) lower .95 upper .95
x1      2.67     0.3745     1.108     6.436

Concordance= 0.64  (se = 0.061 )
Rsquare= 0.156   (max possible= 0.987 )
Likelihood ratio test= 5.08  on 1 df,   p=0.02415
Wald test            = 4.79  on 1 df,   p=0.02867
Score (logrank) test = 5.13  on 1 df,   p=0.02355

> names(fit1);
 [1] "coefficients"     "var"              "loglik"
 [4] "score"            "iter"             "linear.predictors"
 [7] "residuals"        "means"            "concordance"
[10] "method"           "n"                "nevent"
[13] "terms"            "assign"           "wald.test"
[16] "y"                "formula"          "call"
> fit1$loglik;
[1] -65.32214 -62.78010
> fit2<-coxph(Surv(time,status==1)~x2);
> summary(fit2);
```

```
Call:
coxph(formula = Surv(time, status == 1) ~ x2)

  n= 30, number of events= 23

     coef exp(coef) se(coef)     z Pr(>|z|)
x2 0.1884    1.2073   0.4314 0.437    0.662

   exp(coef) exp(-coef) lower .95 upper .95
x2     1.207     0.8283    0.5183     2.812

Concordance= 0.506  (se = 0.061 )
Rsquare= 0.006   (max possible= 0.987 )
Likelihood ratio test= 0.19  on 1 df,   p=0.6605
Wald test            = 0.19  on 1 df,   p=0.6624
Score (logrank) test = 0.19  on 1 df,   p=0.6619

> fit2$loglik;
[1] -65.32214 -65.22568
> fit3<- coxph(Surv(time,status==1)~x1+x2);
> summary(fit3);
Call:
coxph(formula = Surv(time, status == 1) ~ x1 + x2)

  n= 30, number of events= 23

     coef exp(coef) se(coef)     z Pr(>|z|)
x1 1.0463    2.8472   0.4581 2.284   0.0224 *
x2 0.3586    1.4313   0.4401 0.815   0.4152
---
Signif. codes: 0 *** 0.001 ** 0.01 * 0.05 . 0.1   1

   exp(coef) exp(-coef) lower .95 upper .95
x1     2.847     0.3512    1.1600     6.988
x2     1.431     0.6987    0.6041     3.391

Concordance= 0.648  (se = 0.068 )
Rsquare= 0.175   (max possible= 0.987 )
Likelihood ratio test= 5.76  on 2 df,   p=0.05607
```

```
Wald test              = 5.4  on 2 df,   p=0.06718
Score (logrank) test = 5.77  on 2 df,   p=0.05594

> fit3$loglik;
[1] -65.32214 -62.44106
```

Download and load the following packages:
survival, eha, expm, Matrix, msm, mvtnorm, parfm.

Illustration 8.6

```
> library(parfm);
> mydata<-kidney[,-7];
> head(mydata);
  id time status age sex disease
1  1    8      1  28   1   Other
2  1   16      1  28   1   Other
3  2   23      1  48   2      GN
4  2   13      0  48   2      GN
5  3   22      1  32   1   Other
6  3   28      1  32   1   Other
> dim(mydata) ;
[1] 76  6
> names(mydata) ;
[1] "id"     "time"    "status"  "age"    "sex"    "disease"
># We record the sex as 0-1 indicator for ease of computation;
> mydata$sex<-mydata$sex-1;
> mydata$sex; # Data are not displayed;
> #consider the model with the positive stable frailty ;
> # distribution("possta")with exponential baseline;
> # hazard function;
> parfm(Surv(time, status)~ sex+ age, cluster="id",
  data=mydata, dist="exponential", frailty= "possta");

Execution time: 0.88 second(s)

Frailty distribution: positive stable
Baseline hazard distribution: Exponential
Loglikelihood: -337.132
```

```
       ESTIMATE SE p-val
nu       0.000
lambda   0.012
sex     -0.885
age      0.004
---
Signif. codes:0 '***' 0.001 '**' 0.01 '*' 0.05 '.' 0.1 ' '1

Kendall's Tau: 0

> parfm(Surv(time,status) ~ sex+age,cluster="id", data=mydata,
  dist="exponential",frailty="possta",method = "Nelder-Mead");

Execution time: 1.12 second(s)

Frailty distribution: positive stable
Baseline hazard distribution: Exponential
Loglikelihood: -336.182

       ESTIMATE SE    p-val
nu       0.112  0.084
lambda   0.014  0.008
sex     -0.951  0.348 0.006 **
age      0.004  0.011 0.694
---
Signif. codes:0 '***' 0.001 '**' 0.01 '*' 0.05 '.' 0.1 ' '1

Kendall's Tau: 0.112

> #consider the model with the inverse Gaussian distributed;
> #frailties("ingau")with "weibull" baseline hazard function;
> library(parfm);
> mydata<-kidney[,-7];
> head(mydata);
  id time status age sex disease
1  1    8      1  28   1   Other
2  1   16      1  28   1   Other
3  2   23      1  48   2      GN
```

```
4  2   13       0  48   2       GN
5  3   22       1  32   1    Other
6  3   28       1  32   1    Other
> #consider the model with the inverse Gaussian distributed;
> # frailties (ingau);
> # with weibull baseline hazard function;
> library(parfm);
> head(mydata);
  id time status age sex disease
1  1    8      1  28   1   Other
2  1   16      1  28   1   Other
3  2   23      1  48   2      GN
4  2   13      0  48   2      GN
5  3   22      1  32   1   Other
6  3   28      1  32   1   Other
> mydata$sex<-mydata$sex-1;
> mydata$sex; # Data are not displayed;
> parfm(Surv(time, status) ~ sex+ age, cluster="id",
  data=mydata,  dist="weibull", frailty= "ingau");

Execution time: 1.1 second(s)

Frailty distribution: inverse Gaussian
Baseline hazard distribution: Weibull
Loglikelihood: -333.314

       ESTIMATE SE    p-val
theta   0.677   0.540
rho     1.145   0.148
lambda  0.013   0.011
sex    -1.481   0.432 0.001 ***
age     0.006   0.012 0.655
---
Signif. codes: 0 '***' 0.001 '**' 0.01 '*' 0.05 '.' 0.1' '1

Kendall's Tau: 0.181
> library(parfm);
> mydata<-kidney[,-7];
> mydata$sex<-mydata$sex-1;
```

```
> mydata$sex; # Data are not displayed;
> mydata.select<-select.parfm(Surv(time,status) ~ sex + age,
  cluster = "id", data = mydata, dist = c("exponential",
  "weibull"), frailty=c( "ingau", "possta"));


### - Parametric frailty models - ###
Progress status:
  'ok' = converged
  'nc' = not converged

                Frailty
Baseline             invGau  posSta
exponential.........ok......ok....
Weibull.............ok......ok....
> mydata.select;

AIC:
              ingau   possta
exponential 675.699 682.264
weibull     676.627 682.315


BIC:
              ingau   possta
exponential 685.022 691.587
weibull     688.281 693.969
```

Illustration 8.7

```
> library(survival);
> options(warn=-1);
> coxph(Surv(time,status) ~ sex+age +
  frailty(id, distrubution= "gamma", eps=1e-11),
  outer.max=15,data=kidney);

Call:
coxph(formula = Surv(time, status) ~ sex + age + frailty(id,
```

```
   distrubution="gamma",eps =1e-11),data=kidney,outer.max=15)

                        coef     se(coef) se2    Chisq DF   p
sex                     -1.58323 0.4594   0.3515 11.88  1.0 0.00057
age                      0.00522 0.0119   0.0088  0.19  1.0 0.66000
frailty(id, distrubution                         22.96 12.9 0.04100

Iterations: 10 outer, 72 Newton-Raphson
     Variance of random effect= 0.408   I-likelihood = -181.6
Degrees of freedom for terms=  0.6  0.6 12.9
Likelihood ratio test=46.5  on 14.1 df, p=2.37e-05  n= 76
```

R-commands for Illustration 8.8

```
> library(survival);
> library(KMsurv);
> data(larynx);
> attach(larynx);

> head( larynx);
  stage time age diagyr delta
1     1  0.6  77     76     1
2     1  1.3  53     71     1
3     1  2.4  45     71     1
4     1  2.5  57     78     0
5     1  3.2  58     74     1
6     1  3.2  51     77     0
> sr.fit<-survreg(Surv(time,delta)~
  as.factor(stage) + age,dist="weibull");

> summary(sr.fit);

Call:
survreg(formula = Surv(time,delta) ~ as.factor(stage) + age,
    dist = "weibull")
                     Value Std. Error      z        p
(Intercept)         3.5288     0.9041  3.903 9.50e-05
as.factor(stage)2 -0.1477     0.4076 -0.362 7.17e-01
as.factor(stage)3 -0.5866     0.3199 -1.833 6.68e-02
```

```
as.factor(stage)4 -1.5441     0.3633 -4.251 2.13e-05
age               -0.0175     0.0128 -1.367 1.72e-01
Log(scale)        -0.1223     0.1225 -0.999 3.18e-01

Scale= 0.885

Weibull distribution
Loglik(model)= -141.4   Loglik(intercept only)= -151.1
        Chisq= 19.37 on 4 degrees of freedom, p= 0.00066
Number of Newton-Raphson Iterations: 5
n= 90

> # The AIC of the exponential is 285.8;
> # AIC for exponential is lower than AIC for weibull;
```

Exponential distribution is more appropriate.

Exponential model is a special case of Weibull. In part 1, the p-value for Log (scale) is 0.318, which is not significant. This also suggests that exponential model is more appropriate.

Chapter 9

Analysis of Competing Risks

9.1 Introduction

In many situations there are several possible risks or modes of failure. The unique actual risk which claims the life of the unit is called the cause of failure. The risks are said to compete for the life of the unit, hence the probabilistic model for the lifetime in the presence of several risks is called the competing risks model.

The study of the competing risks model has a fascinating history. It goes as far back as at least D. Bernoulli (1760) when he studied the effect of eliminating small pox as a possible risk on the life expectancy of humans.

There is a close correspondence between the competing risks model and the series system of components. Recall that the series system operates only as long as all the components operate and fails as soon as anyone of the components fails. Hence one may envisage the components as risks competing for the life of the system.

Actually, this model has been seen to be useful in much wider circumstances, which are described as 'time-event' situations wherein an event takes place at a random time and results in one of a finite number of possible outcomes. In competing risks or the series system, the event is the failure of the unit and the outcome is the cause of failure or the component which fails. In other situations the event could be as diverse as (i) dissolution of a marriage with the duration of marriage as the random time and the mode of dissolution (death of a spouse / divorce) as the outcome or (ii) period of unemployment, type of employment secured, etc. The methodology developed for competing risks is thus applicable in many disciplines such as economics, sociology, etc., besides engineering and medical studies.

9.2 The Model for General Competing Risks

In all the situations described above it is to be noted that the observation consists of a positive valued continuous random variable T indicating the time at which the event (failure, death, etc.) takes place and the outcome of the event (cause of death, indicator of the component which failed, etc.) which is a discrete random variable δ, taking values $1, 2, \cdots, k$; assuming there are k possible outcomes, risks, components, etc. Therefore, we need to model the probabilistic behaviour of the random pair (T, δ). If we have n independent units then the data will consist of n independent pairs $(T_i, \delta_i), i = 1, 2, \cdots, n$.

The joint distribution of (T, δ) may be specified in terms of the k sub-survival functions

$$\overline{F}(t, j) = P(T > t, \delta = j), j = 1, 2, \cdots, k.$$

One can also define the subdensity functions $f(t, j) = -\frac{\partial}{\partial t}\overline{F}(t, j)$. The proper density of T is then $f(t) = \sum_{j=1}^{k} f(t, j)$. Similarly, the marginal probability distribution of δ is given by $P(\delta = j) = \overline{F}(0, j) = p_j$, say. Then $\sum_{j=1}^{k} p_j = 1$.

The overall hazard rate of T is $h(t) = -\frac{\partial}{\partial t}\log\overline{F}(t) = f(t)/\overline{F}(t)$ and the cause specific hazard rate is defined as $h(t, j) = \frac{f(t,j)}{\overline{F}(t)}$ leading to the relationship $h(t) = \sum_{j=1}^{k} h(t, j)$. The cause specific hazard rates are also known as crude hazard rates. The relative risk of the j-th mode (j-th competing risk) is defined by the ratio $h(t, j)/h(t)$. It is seen that this ratio is constant (independent of t) if and only if T and δ are independent. The 'constant relative risk' phenomenon is also known as 'proportional hazards' in the context of competing risks. This can hold even when the risks are not independent.

We may also define the conditional survival function of the j-th cause of failure as

$$\overline{F}(t|j) = P(T > t|\delta = j] = \frac{P[T > t, \delta = j]}{P(\delta = j]} = \frac{\overline{F}(t, j)}{p_j}.$$

Therefore in case T and δ are independent, $\overline{F}(t) = \overline{F}(t|j)$, for every j.

In general, one may express

$$\overline{F}(t) = \sum_{j=1}^{k} p_j \ \overline{F}(t|j)$$

which is a mixture of the conditional survival function $\overline{F}(t|j)$ with weights p_j.

This representation leads to mixture models based on well-known distributions such as exponential or Weibull for $\overline{F}(t|j)$.

9.3 Independent Competing Risks

Substantially simpler models can be provided for the competing risks situation if it can be assumed 'a priori', that the risks operate independently of each other. For the human population it has been argued that the following four groups of risks, viz. (i) Cardiovascular disease, (ii) neoplasms, (iii) accidents and violence, and (iv) all others, act independently. As will be demonstrated later, the data on (T, δ) generated in the competing risks situation is not useful for distinguishing between dependent and independent risks. Hence, the assumption of independence has to be based on knowledge outside of the data.

A common way of formulating a model is to define $X_1, X_2, \cdots, X_k$ respectively as the 'latent' lifetimes of the unit when it is exposed to the $1^{st}, 2^{nd}, \cdots, k^{th}$ risk alone. Then $T = \min(X_1, X_2, \cdots, X_k)$ and $\{\delta = j\} \equiv \{T = X_j\}$.

Let $\overline{F}(t_1, t_2, \cdots, t_k)$ be the joint survival function of $X_1, X_2, \cdots, X_k$. The marginal survival function of T is then

$$\overline{F}(t) = \overline{F}(t, \cdots, t).$$

The joint probability distribution of T and δ is given by the subdensity functions.

$$f(t, j) = \left.\frac{-\partial \overline{F}(t_1, \cdots, t_k)}{\partial t_j}\right|_{t_1 = \cdots = t_k = t}.$$

Or the associated subsurvival functions

$$\overline{F}(t, j) = \int_t^\infty f(u, j) du.$$

However, this kind of identifying relationship does not exist in the reverse direction. In other words for a given set of $\overline{F}(t,j), j = 1, \cdots, k$, there will exist a unique joint distribution of $F(t_1, \cdots, t_k)$ of $X_1, \cdots, X_n$ only if these are independent. Otherwise, an infinity of joint distributions of $X_1, \cdots, X_n$ are consistent with a given joint distribution of (T, δ). This is the 'non-identifiability' inherent in the latent lifetimes model for competing risks.

(a) : If $X_1, \cdots, X_k$ are independent and also identically distributed then

$$\begin{aligned} P(T > t) = \overline{F}(t) &= P(X_1 > t, \cdots X_k > t] \\ &= \prod_{j=1}^{k} P(X_j > t) = \prod_{j=1}^{k} \overline{F}_j(t) = (\overline{G}(t))^k \\ &\qquad \text{if} \quad \overline{F}_j(t) = \overline{G}(t) \ \ \forall \ j. \end{aligned}$$

Therefore, $\overline{G}(t) = (\overline{F}(t))^{1/k}$, showing that the marginal distribution of X_j are uniquely determined by the survival function of T, the time to failure.

(b): If $X_1, X_2, \cdots, X_k$ are independent but not identically distributed and have marginal survival functions $\overline{G}_j(t) = P(X_i > t)$, then one can see that

$$\overline{G}_j(x) = \exp \int_x^{\infty} \left[1 - \sum_{i=1}^{k} F(t,i) \right]^{-1} dF(t,j), j = 1, \cdots, k.$$

Thus showing that the joint survival function of $X_1, \cdots, X_k$, viz.

$$\overline{F}(t_1, \cdots, t_k) = \prod_{j=1}^{k} \overline{G}_j(t_j)$$

is identifiable from the joint distribution of (T, δ).

(c): The $\overline{G}_j(t)$ defined above are obtainable uniquely from any set of $\overline{F}(t,j)$ leading to $\prod_{j=1}^{k} \overline{G}_j(t_j)$ as the unique independent latent lifetimes model consistent with the given $\overline{F}(t,j)$. However, an infinite number of dependent models for the joint distribution of $(X_1, \cdots, X_k)$ are consistent with this set of $\overline{F}(t,j)$.

9.4 Bounds on the Joint Survival Function

It is easily seen that

$$\overline{F}(\max(t_1, \cdots, t_k)) \leq \overline{F}(t_1, \cdots, t_k) \leq \sum_{j=1}^{k} \overline{F}(t_j, j)$$

since the middle term is the probability of the set $\left[\prod_{i=1}^{k}(t_i, \infty)\right]$ and the lower and upper bounds are probabilities of smaller and larger sets respectively. These are the best bounds for the joint survival functions given the subsurvival functions without restricting the generality of the results.

9.5 The Likelihood for Parametric Models with Independent Latent Lifetimes

Suppose n identical experimental units are subject to failure by any of the k competing risks.

Let us now look at the likelihood of the observations (T_i, δ_{ij}), where $T_i = \min(X_{i1}, \cdots, X_{ik})$ and $\delta_{ij} = I[T_i = X_{ij}), i = 1, 2, \cdots, n$.

Let us further assume that the latent lifetimes are independent exponential random variables with mean $\theta_j, j = 1, 2, \cdots, k$ respectively. Then

$$\begin{aligned} L &= \prod_{j=1}^{k}\prod_{i=1}^{n}\{g_j(t_i)\}^{\delta_{ij}}[\overline{G}_j(t_i)]^{1-\delta_{ij}} \\ &= \prod_{j=1}^{k} L_j, \\ g_j(t_i) &= \frac{1}{\theta_j}e^{-t_i/\theta_j}, \overline{G}_j(t_i) = e^{-t_i/\theta_j}, \theta_j > 0, t_i > 0 \\ & \qquad\qquad i = 1, \cdots, n, \quad j = 1, \cdots, k. \end{aligned}$$

Then

$$\begin{aligned} L &= \prod_{j=1}^{k}\prod_{i=1}^{n}(\frac{1}{\theta_j}e^{-t_i/\theta_j})^{\delta_{ij}}(e^{-t_i/\theta_j})^{(1-\delta_{ij})} \text{ and} \\ L_j &= \left(\frac{1}{\theta_j}\right)^{\Sigma_i \delta_{ij}} \exp\{\sum_{i=1}^{n}\frac{t_i}{\theta_j}\}. \end{aligned}$$

Let $\Sigma_i \delta_{ij} = n_j$ and $\sum_{i=1}^{n} t_i = t$. Then the likelihood equation for θ_j is $\log L_j = -n_j \log \theta_j - \frac{t}{\theta_j}$ giving its maximum likelihood estimator $\hat{\theta}_j = \frac{t}{n_j}, j = 1, \cdots, k, n_j > 0$. In case $n_j = 0$ for any j then it is not possible to estimate the corresponding θ_j since no failure due to the j-th risk has been observed. If we denote the failure rate by λ_j then $\lambda_j = \frac{1}{\theta_j}$ and $\hat{\lambda}_j = \frac{n_j}{t}$ by invariance property of MLE's.

However, $\hat{\lambda}_j$ is not unbiased since (letting n_j and t represent the corresponding random variable also)

$$E(\hat{\lambda}_j) = E(n_j) \cdot E(1/t) = \frac{n}{n-1} \lambda_j.$$

But the small modification $\hat{\lambda}'_j = \frac{n-1}{n} \cdot \frac{n_j}{t}$ provides an unbiased estimator for λ_j.

The variance of this unbiased estimator, for $n > 2$ is $V(\hat{\lambda}'_j) = [(n-1)\Delta\lambda_j + \lambda_j^2]/n$ where $\Delta = \Sigma \frac{1}{\theta_j} = \Sigma \lambda_j$. And the covariances are given by

$$Cov(\hat{\lambda}'_j, \hat{\lambda}'_{j'}) = \frac{\lambda_j \, \lambda_{j'}}{n(n-2)}.$$

We may replace the unknown λ_j in these expressions by their maximum likelihood estimates to get the estimated variance-covariance matrix. The elements of this matrix, can then be used to construct approximate confidence intervals based on the asymptotic normality of the estimators.

It is difficult to obtain explicit solutions to the likelihood equations in other cases, although numerical solutions can be found by using iterative schemes such as the Newton-Raphson technique. For details refer to David and Moeschberger (1978) and Crowder (2001).

9.6 Tests for Stochastic Dominance of Independent Competing Risks

(a) Likelihood based approach

In case we do not have a preference for a specific parametric model, one can go for distribution-free methods by concentrating on the ranks of the lifetimes and the indicators of their cause of death.

The likelihood of the ranks and the indicators are derived as follows:

From now on let us specialise to $k = 2$. That is, we simplify the situation to that where only two risks are operating. So the data consists of $T_1, T_2, \cdots, T_n$, the lifetimes of the n units and $\delta_i = I(X_{1i} > X_{2i})$, the indicator of the event that the second risk claimed the life. Let F and G be the distribution functions of X_1 and X_2 respectively. We parametrise the problem to some extent by assuming that F and G belong to the same parametric family and $F = F_0$ and $G = F_\theta$, for some fixed $\theta > 0$. As is usual in non-parametric inference now we pose the following testing problem:

$H_0 : F_0(x) = F_\theta(x)$ vs. $H_1 : F_0(x) \leq F_\theta(x)$ with strict inequality for some x.

The H_1 indicates that the second risk is more likely to claim the life of the unit than the first, upto any age x. We define $T_{(1)}, T_{(2)}, \cdots, T_{(n)}$ to be the ordered lifetimes and

$$W_i = \begin{cases} 1, \text{ if } T_{(i)} \text{ corresponds to a lifetime claimed by the second risk} \\ 0, \text{ otherwise.} \end{cases}$$

Besides, we have $R_i =$ Rank of T_i among $T_1, T_2, \cdots, T_n$. Then $(\underline{R}, \underline{w}) = (R_1, R_2, \cdots, R_n, w_1, w_2, \cdots, w_n)$ constitutes the rank data in this setup.

It is easily seen that $(\underline{R}, \underline{w})$ takes $n!2^n$ possible values given by the permutations of $(1, 2, \cdots, n)$ for $R_1, \cdots, R_n$ and each w_i being either 0 or 1.

We can also see that

$$P(\underline{R}, \underline{w}|H_0) = \frac{1}{n!2^n}$$

and, in general

$$P(\underline{R}, \underline{w}|H_1) = \int_{0<t_1\cdots<t_n<\infty} \cdots \int \prod_{i=1}^{n} [f_\theta(t_i)\overline{F}_0(t_i)]^{w_i} [f_0(t_i)\overline{F}_\theta(t_i)]^{1-w_i} dt_i.$$

The locally most powerful rank (LMPR) size - α test is then specified by a critical region consisting of the union of $M = \alpha \cdot n\ 2^n$ rank sets for which $\frac{\partial p}{\partial t}(\underline{R}, \underline{w}|\theta)\Big|_{\theta=0}$ is largest.

Some calculations lead us to the statistic

$$V = \sum_{j=1}^{n} (a_j W_j - b_j(1 - W_j))$$

where

$$a_j = n!2^n \int_{0<t_1\cdots t_n<\infty} \cdots \int \frac{\overline{f}_0(t_j)}{f_0(t_j)} \prod_{i=1}^{n} f_0(t_i)\overline{F}_0(t_i)dt_i$$

and

$$b_j = n!2^n \int \cdots \int \frac{f_0^*(t_j)}{\hat{F}_0(t_j)} \prod_{i=1}^{n} f_0(t_i)\overline{F}_0(t_i)dt_i dt_c$$

where $\overline{f}_0(t_j) = \left[\frac{\partial f_\theta(a)}{\partial \theta}\right]_{\theta=0}$ and $f_0^*(t_j) = \left[\frac{\partial F_\theta(x)}{\partial \theta}\right]_{\theta=0}$.

Examples

(a) Let $f_\theta(x) = \frac{e^{x+\theta}}{1+e^{x+\theta}}$: logistic distribution. In the complete two-sample situation we know that the Wilcoxon test based on $W = \Sigma R_i$ is the LMPR test. In the present competing risks situation the LMPR test is based on the statistic

$$V = \sum_{j=1}^{n} (1 - c_j) W_j$$

where

$$c_j = \frac{1}{2n+1} + \sum_{k=2}^{j} \frac{2n(2n-2)\cdots(2n-2k+4)}{(2n+1)(2n-1)\cdots(2n-2k+3)}$$

(b) If $F_\theta(x) = 1 - e^{-(1+\theta)x}$, $\theta \geq 0$, then $W = \sum_{j=1}^{n} W_j = \sum_{j=1}^{n} \delta_j$, the sign statistic gives the LMPR test, but this does not use the lifetime $T_1, \cdots, T_n$ or their ranks. This is because $\underline{T}$ and $\underline{\delta}$ are independent both under H_0 and H_1. (Usually they are independent under H_0 but not under H_1.) Hence $P(\underline{T}, \underline{W})|\theta) = P(\underline{T}|\theta)P(\underline{W}|\theta) = \frac{1}{n!}P(\underline{W}|\theta)$. Therefore only the distribution of W changes from H_0 to H_1, leading to the optimal test based only on $\underline{W}$, which is the sign test.

We note that to derive the LMPR test one must know F_θ, the family of probability distributions to which the distributions of the latent lifetimes X_1 and X_2 belong. In many situations this information is not available. Also, even in simple situations the tests may have to be based on rather complicated scores. In the next section we describe certain non-parametric tests, based on simple statistics which have reasonably good properties.

(c) *Tests based on a heuristic principle for non-parametric alternatives*

Earlier we parametrised the alternative by specifying $F = F_0$ and $G = F_\theta$ with $F_0(x) \leq F_\theta(x)$. Now we do not do so and frame the H_0 and H_1 as $H_0 : F(x) = G(x)$ and $H_1 : F(x) \leq G(x)$ again indicating that the second risk is more likely to be the cause of failure than the first upto any age x. We look simultaneously at the pairs (T_i, δ_i) and (T_j, δ_j) and define the kernel function

$$\phi(T_i, \delta_i, T_j, \delta_j) = \begin{cases} 1, \text{ if } \delta_i = 1, \ T_i < T_j \\ 1, \text{ if } \delta_j = 1, \ T_j < T_i \\ 0 \text{ otherwise.} \end{cases}$$

It is seen that $E(\phi) = \frac{1}{2}$ under H_0 and greater than 1/2 under H_1.

Also, it can be seen that $\{\phi = 1\}$ indicates the event $\min(Y_i, Y_j) < \min(X_i, X_j)$ and $\{\phi = 0\}$ indicates the complementary event. It is felt that the probability of $\{\phi = 1\}$ is a functional which is able to discriminate between the H_0 and H_1. We define its U-statistic estimator below.

$$U^* = \frac{1}{\binom{n}{2}} \sum\sum_{i<j} \phi(T_i, \delta_i, T_j, \delta_j)$$

which is equal to

$$\frac{1}{\binom{n}{2}} \sum_{i=1}^{n} (n - R_i)\delta_i.$$

One may consider a slightly different statistic

$$U = \frac{1}{\binom{n}{2}} \sum_{i=1}^{n} (n - R_i + 1)\delta_i$$

to include the number of pairs (X_i, Y_i) in which $X_i < Y_i$. Then the exact distribution under H_0 can be investigated through its moment generating function. If $S = \binom{n}{2} U$, then

$$M_s(t) = 2^{-n} \prod_{j=1}^{n} \{1 + e^{jt}\}.$$

This gives $E_{H_0}(S) = \frac{n(n+1)}{4}$ and $V_{H_0}(S) = \frac{n(n+1)(2n+1)}{24}$. This moment generating function is in fact the same as the null m.g.f. of the Wilcoxon signed rank statistic W^+. Hence, the tables of critical values available for W^+ may be used for S also. One may easily see that asymptotically the difference between U and U^* goes to zero in probability and under H_0,

$$(3n)^{\frac{1}{2}} \left(U^* - \frac{1}{2} \right)$$

has the standard normal distribution, allowing the use of the critical points from the $N(0,1)$ distribution. See Bagai, Deshpande and Kochar (1989) for a detailed discussion.

Illustration 9.1: We consider the well-known Boag's data (1949). The following are the survival times (in months) for 121 breast cancer patients. It comes from the clinical records of one hospital from the years 1929 to 1938.

Cancer: 317, 318, 399, 495, 525, 536, 549, 552, 554, 557, 558, 571, 586, 694, 596, 605, 612, 621, 628, 631, 636, 643, 647, 648, 649, 661, 663, 666, 670, 695, 697, 700, 705, 712, 713, 738, 748, 753, 159, 189, 191, 198, 200, 207, 220, 235, 245, 250, 256, 261, 265, 266, 280, 343, 356, 383, 403, 414, 428, 432.

Others: 0.3, 4.0, 7.4, 15.5, 23.4, 46, 46, 51, 65, 68, 83, 88, 96, 110, 111, 112, 132, 162, 111, 112, 113, 114, 114, 117, 121, 123, 129, 131, 133, 134, 134, 136, 141, 143, 167, 177, 179, 189, 201, 203, 203, 213, 228.

The causes of death are cancer (1) and others (0). We want to test whether deaths due to cancer occur more often than others. Assume that the two risks are independent.

Solution: We have used asymptotic normal distribution of test statistics as the sample size is large. The value of Standardised U^* statistics is 5.849608.

Conclusion: Standardised U^* is significant rejecting the null hypothesis of equality of distributions. Failures due to cancer occur more often.

For R-commands see the appendix of this chapter.

9.7 Tests for Proportionality of Hazard Rates of Independent Competing Risks

Let X_1 and X_2 be the two independent latent lifetimes of the unit under the risks, with marginal survival functions $\overline{F}$ and $\overline{G}$ respectively. Further, let λ_F and λ_G be the corresponding hazard rates, which we wish to compare. We say that the failure of the unit due to the two risks occur at the same relative rate if $H_0 : \frac{\lambda_F(x)}{\lambda_G(x)} = a$ (constant) for every x, is true. Alternatively, we say that the failures due to risk I occur relatively faster, as age increases, if $H_1 : \frac{\lambda_F(x)}{\lambda_G(x)}$ is increasing in x.

We construct a test statistic on heuristic grounds which enables us to discriminate between the above two hypotheses. Consider the ratio of the two failure rates at two points $x_1 < x_2$. It is clear that the difference $\frac{\lambda_F(x_2)}{\lambda_G(x_2)} - \frac{\lambda_F(x_1)}{\lambda_G(x_1)} = 0$ under H_0, and is positive under H_1. Equivalently, one can say that $\lambda_F(x_2)\lambda_G(x_1) \geq \lambda_F(x_1)\lambda_G(x_2)$ under H_1 with the equality holding under H_0 for all $x_1 < x_2$. After some algebra it is seen to be equivalent to

$$\delta(x_1, x_2) = f(x_2)\overline{G}(x_2)\overline{F}(x_1)g(x_1) - f(x_1)\overline{G}(x_1)\overline{F}(x_2)g(x_2) \geq 0 \,\forall\; x_1 < x_2,$$

with equality holding under H_0.

Let us define the real valued parameter

$$\Delta(F, G) = \int_0^\infty \int_0^{x_2} \delta(x_1, x_2)dx_1 dx.$$

It is obvious that under $H_0 \Delta(F, G) = 0$ and under $H_1, \Delta(F, G) > 0$. The double integral can be written as

$$\Delta(F, G) = I_1 - I_2$$

where

$$I_1 = \int_0^\infty \int_0^{x_2} \overline{G}(x_2)\overline{F}(x_1)dF(x_2)dG(x_1)$$

and

$$I_2 = \int_0^\infty \int_0^{x_2} \overline{G}(x_1)\overline{F}(x_2)dF(x_1)dG(x_2).$$

We can interpret these as

$$I_1 = P(Y_1 < X_2 < Y_2, Y_1 < X_1]$$

and

$$I_2 = P(X_1 < Y_2 < X_2, X_1 < Y_1],$$

where X_1 and X_2 are two independent random variables with common c.d.f. F and Y_1 and Y_2 are further independent random variables with common c.d.f. G.

The data calculated from n units which have failed is $(T_i, \delta_i), i = 1, 2, \cdots, n$ consisting of the lifetimes of the units and the indicator of the first risk being the cause of failure. In other words, $T_i = \min(X_i, Y_i)$ and

$\delta_i = I(X_i < Y_i]$. We define a kernel

$$\psi(T_1, T_2, \delta_1, \delta_2) = \begin{cases} 1, & \text{if } \delta_2 = 1,\, \delta_1 = 0,\, T_1 < T_2 \\ & \text{or } \delta_1 = 1,\, \delta_2 = 0,\, T_2 < T_1 \\ -1, & \text{if } \delta_2 = 1,\, \delta_1 = 0,\, T_2 < T_1 \\ & \text{or } \delta_1 = 1,\, \delta_2 = 0,\, T_1 < T_2 \\ 0, & \quad \text{otherwise.} \end{cases}$$

Then it is seen that

$$E(\psi) = 2(I_1 - I_2) = 2\Delta(F, G)$$

which is 0 under H_0 and strictly positive under H_1. Now construct the U-statistic based on the kernel ψ:

$$V = \frac{1}{\binom{n}{2}} \sum\sum_{1 \leq i < j \leq n} \psi(T_i, T_j, \delta_i, \delta_j).$$

Then a heuristic test for H_0 vs. H_1 is : Reject H_0 if V is too large.

A simpler version of V may be seen to be in terms of $W_1, W_2, \cdots, W_n$, where $W_i = 1$, if the order statistic $T_{(i)}$ of $T_1, T_2, \cdots, T_n$ is an X observation and $= 0$ if it is a Y observation. Then

$$V = \frac{1}{\binom{n}{2}} S = \frac{1}{\binom{n}{2}} \left[2\sum_{i=1}^{n} iW_i - (n+1)\sum_{i=1}^{n} W_i \right].$$

Here it may be noted that $\sum_{i=1}^{n} iW_i$ and $\sum_{i=1}^{n} W_i$ are the Wilcoxon signed rank statistic and the sign statistic adapted to the competing risks data. Under H_0, the moment generating function of S, and hence its mean and variance can be easily obtained since $W_1, W_2, \cdots, W_n$ are independent with distribution $P[W_i = 1] = \frac{a}{a+1}$ and $P(W_i = 0] = \frac{1}{a+1}$. Using this moment generating function of S is

$$M_S(t) = \prod_{i=1}^{n} \left[\frac{1}{a+1} + \frac{a}{a+1} \exp\{(n+1-2i)t\} \right].$$

From this moment generating function, it is easy to obtain moment generating function of V, which is scaled version of S and its mean and variance under H_0.

This gives $E_{H_0}(V) = 0$ and $E_{H_0}(V^2) = \frac{4}{3}\frac{(n+1)}{n(n-1)}\frac{a}{(a+1)^2}$. Giving the asymptotic value

$$\lim_{n\to\infty} n^{1/2} V_{H_0}(V) = \frac{4}{3}\frac{a}{(a+1)^2}.$$

Let $U_s = \frac{1}{n}\sum_{i=1}^{n} W_i$ be the sign statistic. It is easily seen that $U_s(1-U_s)^{-1}$ is a consistent estimator of the unknown parameter a. Thus by Slutsky's theorem and the U-statistics limit theorem one concludes that $V^* = \{\frac{4}{3}U_s(1-U_s)\}^{-1/2} n^{1/2} V$ tends in distribution to the standard normal random variable. Thus the asymptotic critical points for carrying out the test are provided by the $N(0,1)$ distribution. See Deshpande and Sengupta (1995) for further details.

Illustration 9.2: Use the data of illustration 9.1. Test the hypothesis of proportionality of hazard rates assuming that the two risks are independent.
Solution: We have used asymptotic normal distribution of test statistics as the sample size is large. The value of Standardised V^* statistics is 6.486423.
Conclusion: Standardised V^* is significant rejecting the null hypothesis. So the ratio of failure rates is not proportional.
For R-commands see the appendix of this chapter.

9.8 Tests in the Context of Dependent Competing Risks

(a) Tests for Equality of the Incidence Functions (Aras and Deshpande (1992)).

As before, suppose that a unit is exposed to competing risks denoted by 1 and 2. The data, when n units are put on trial consists of $(T_i, \delta_i), i = 1, 2, \cdots, n$ where T_i is the lifetime of the unit and $\delta_i = 1, 2$ according to the risk which claims the life. Since the two risks are not assumed to be independent, the probability distribution of (T, δ) given by the two subdistribution functions $F(1,t) = P(T \le t, \delta = 1)$ and $F(2,t) = P(T \le t, \delta = 2)$ does not identify the joint distribution of (X, Y) the latent lifetimes of the unit under the two risks. The hypotheses that the two risks are equally effective (forceful) at all ages, is then described by

$$H_0 : F(1,t) = F(2,t) \ \text{ at all } \ t$$

or equivalently $H_0 : f(1,t) = f(2,t)$, where $f(i,t)$ are the subdensity functions corresponding to $F(i,t)$. The likelihood function of the data is given by

$$L(\underline{T}, \underline{\delta}, f(1,t_i), f(2,t_i)) = \prod_{i=1}^{n} (f(1,t_i))^{\delta_i^*} [f(2,t_i)]^{1-\delta_i^*}$$

where $\delta^* = 2 - \delta$.

In models leading to independence of T and δ, for example where $F(1,t) = \theta H(t)$ and $f(2,t) = (1-\theta)H(t), 0 \leq \theta \leq 1$, with H being a distribution function, the null hypothesis and alternative reduce to $H_0 : \theta = \frac{1}{2}$ and $H_1 : \theta \neq \frac{1}{2}$. It is easily seen that the optimal test is then based on the sign statistic

$$U_1 = \frac{1}{n} \sum_{i=1}^{n} \delta_i^*.$$

However, the more general situation is described by $f(1,t) = f(1,t,\theta)$ and $f(2,t) = h(t) - f(1,t,\theta)$ where $h(t)$ and $f(1,t,\theta)$ are known, the null hypothesis being provided by $f(t,\theta_0) = \frac{1}{2}h(t)$. Now if we choose $h(t,\theta) = e^{(x-\theta)}[1+e^{(x-\theta)}]^{-2}$, the logistic density, then the LMP rank test is based on the Wilcoxon signed rank like statistic

$$W^* = \sum_{i=1}^{n} iW_i.$$

Here W_i is the value of δ^* corresponding to the T_j with rank i.

In case the model is given by $F(1,t) = \left(\frac{H(t)}{\theta}\right)^2$ and $F(2,t) = H(t) - \left(\frac{H(t)}{2}\right)^{\theta}$, the LMPR test is based on the statistic $L_c = \sum_{i=1}^{n} W_i a_i$ where $a_i = E(X_{(i)})$, and $X_{(i)}$ is the i-th order statistic from a random sample of size n from the standard exponential distribution.

(b) Tests for Bivariate Symmetry

Assume that the latent failure times X and Y are dependent with the joint c.d.f. $F(x,y)$. Then on the basis of the data $(T_i, \delta_i^*), i = 1, 2, \cdots, n$ one may wish to test the null hypothesis of bivariate symmetry, viz.,

$$H_0 : F(x,y) = F(y,x) \quad \text{for all } x \text{ and } y.$$

Under this null hypothesis we have (i) $F(1,t) = F(2,t)$ for all t, (ii) $\lambda(1,t) = \lambda(2,t)$ for all t, (iii) $P(\delta^* = 1) = P(\delta^* = 0)$ and also (iv) T and δ^* are independent. In view of these observations Kochar and Carriere (2000) have suggested tests which are useful to detect the alternatives

$$H_1 : \lambda(1,t) < \lambda(2,t) \;\; \forall \;\; t$$

$$H_2 : F(1,t) < F(2,t) \;\; \forall \;\; t.$$

They have suggested the use of $U_2 = \sum_{i=1}^{n-1} i(n-i)W_{i+1}$ for testing H_0 against H_1. They have shown that

$$n^{1/2}\left[\frac{U_2}{n^3} - \frac{1}{12}\right] \xrightarrow{d} N(0, \frac{1}{120})$$

which would provide approximate critical points in case the sample size n is at least moderately large. They have also provided the exact critical points for $5 \leq n \leq 20$. Further, the other tests proposed earlier in this chapter may also be selectively used for detecting the suspected alternative hypothesis.

Exercises

Exercise 9.1: Hoel's data (1972, Biometrics, 28, 475-488) arise from the a laboratory experiment in which male mice of strain RFM were given radiation dose 300 rads at 5 to 6 weeks old. There are two groups of mice: conventional lab environment (group 1) and germ-free environment (group 2). The survival times are measured in days and the causes of death are (1) thymic lymphoma, (2) reticulum cell sarcoma, and (3) others. We consider the data for group 1.

Survival times of mice (Group 1)

Cause 1: 150, 189, 191, 198, 200, 207, 220, 235, 245, 250, 256, 261, 265, 266, 280, 343, 356, 383, 403, 414, 428, 432.
Cause 2: 317, 318, 399, 495, 525, 536, 549, 552, 554, 557, 558, 571, 586, 594, 596, 605, 612, 621, 628, 631, 636, 643, 647, 648, 649, 661, 663, 666, 670, 695, 697, 700, 705, 712, 713, 738, 748, 753.
Cause 3: 40, 42, 51, 62, 163,179, 206, 222, 228, 252, 259, 282, 324, 333, 341, 366, 385, 407, 420, 431, 441, 461, 462, 482, 517, 517, 524, 564, 567,586, 619, 620, 621, 622, 647, 651, 686, 761, 763.

(A): Consider the causes of death as cancer (1) and others (0) and test whether deaths due to cancer occur more often than others. Assume independence of the two risks.
(B): Consider the causes of death as reticulum cell sarcoma (1) and others (0) and test whether deaths due to reticulum cell sarcoma occur more often than Others. Assume independence of the two risks.
Exercise 9.2: Consider the data of Exercise 9.1 (A). Assume independence of the two risks. Carry out test for proportionality of hazards of the two risks.

Appendix

Test for stochastic dominance of competing risks: Illustration 9.1.

```
> Cancer<-c(0.3, 5.0, 5.6, 6.2, 6.3, 6.6, 6.8, 7.5, 8.4,
  8.4,  10.3, 11.0, 11.8,  12.2, 12.3, 13.5, 14.4, 14.4,
  14.8, 15.7, 16.2, 16.3,  16.5, 16.8, 17.2, 17.3, 17.5,
  17.9, 19.8, 20.4, 29.9,  21.0, 21.0, 21.1, 23.0, 23.6,
  24.0, 24.0, 27.9, 28.2, 29.1, 30.0, 31, 31, 32, 35,
  35, 38, 39, 40, 40, 41, 41, 42, 44, 46, 48, 48, 51,
  51, 52, 54, 56, 60, 78, 78, 80, 84, 87, 69, 90, 97,
  98, 100, 114, 126, 131, 174);

> length(Cancer)->n1;

> Others<- c(0.3, 4.0, 7.4, 15.5, 23.4, 46, 46, 51, 65, 68,
  83, 88, 96, 110, 111, 112, 132, 162, 111, 112, 113, 114,
  114, 117, 121, 123, 129, 131, 133, 134, 134, 136, 141,
  143, 167, 177, 179, 189, 201, 203, 203, 213, 228);
> length(Others)->n2;

> d1<-rep(1,n1); d2<-rep(0,n2);

> t<-c(Cancer,Others);

> n<-length(t); n;
[1] 121

> id<-c(d1,d2);  #Identifier of risks;
```

```
> length(id);
[1] 121

> r<-rank(t)#vector of assigned ranks;
> length(r);
[1] 121

> u<-(2/(n*(n-1)))* sum((n-r)*id)
> #computation of test statistics; u;

> sqrt(n)*(u-0.5)/sqrt(1/3)->zo;
> # asymptotically normal test statistics;
> zo;
[1] 5.849608
```

Test for proportionality of hazards: Illustration 9.2

```
> Cancer<-c(0.3, 5.0, 5.6, 6.2, 6.3, 6.6, 6.8, 7.5, 8.4, 8.4,
  10.3, 11.0, 11.8, 12.2, 12.3, 13.5, 14.4, 14.4, 14.8, 15.7,
  16.2, 16.3, 16.5, 16.8, 17.2, 17.3, 17.5, 17.9, 19.8, 20.4,
  29.9, 21.0, 21.0, 21.1, 23.0, 23.6, 24.0, 24.0, 27.9, 28.2,
  29.1, 30.0, 31, 31, 32, 35, 35, 38, 39, 40, 40, 41, 41, 42,
  44, 46, 48, 48, 51, 51, 52, 54, 56, 60, 78, 78, 80, 84, 87,
  69, 90, 97, 98, 100, 114, 126, 131, 174);

> length(Cancer)->n1;

> Others<- c(0.3, 4.0, 7.4, 15.5, 23.4, 46, 46, 51, 65, 68,
  83, 88, 96, 110, 111, 112, 132, 162, 111, 112, 113, 114,
  114, 117, 121, 123, 129, 131, 133, 134, 134, 136, 141,
  143, 167, 177, 179, 189, 201, 203, 203, 213, 228);
> length(Others)->n2;

> d1<-rep(1,n1); d2<-rep(0,n2);
> t<-c(Cancer,Others);
> n<-length(t);  n;
[1] 121
```

```
> id<-c(d1,d2);

> length(id);   t;
[1] 121
  [1]   0.3   5.0   5.6   6.2   6.3   6.6   6.8   7.5   8.4
 [10]   8.4  10.3  11.0  11.8  12.2  12.3  13.5  14.4  14.4
 [19]  14.8  15.7  16.2  16.3  16.5  16.8  17.2  17.3  17.5
 [28]  17.9  19.8  20.4  29.9  21.0  21.0  21.1  23.0  23.6
 [37]  24.0  24.0  27.9  28.2  29.1  30.0  31.0  31.0  32.0
 [46]  35.0  35.0  38.0  39.0  40.0  40.0  41.0  41.0  42.0
 [55]  44.0  46.0  48.0  48.0  51.0  51.0  52.0  54.0  56.0
 [64]  60.0  78.0  78.0  80.0  84.0  87.0  69.0  90.0  97.0
 [73]  98.0 100.0 114.0 126.0 131.0 174.0   0.3   4.0   7.4
 [82]  15.5  23.4  46.0  46.0  51.0  65.0  68.0  83.0  88.0
 [91]  96.0 110.0 111.0 112.0 132.0 162.0 111.0 112.0 113.0
[100] 114.0 114.0 117.0 121.0 123.0 129.0 131.0 133.0 134.0
[109] 134.0 136.0 141.0 143.0 167.0 177.0 179.0 189.0 201.0
[118] 203.0 203.0 213.0 228.0

> timeid<-data.frame(t,id);
> timeid[order(t),];        # output is not shown;

> d3<- timeid[order(t),];
> head(d3);      # Output is not shown;

> I=seq(1:n);

> S3<-sum((n+1-2*I)*d3$id);

> U3<-S3*(2/(n*(n-1))); U3;
[1] 0.3258953

> Us<-1/n*sum(id);
> V0<-((4/3)*Us*(1-Us))^-.5*(sqrt(n)*U3);
> V0;
[1] 6.486423
```

Chapter 10

Repairable Systems

10.1 Introduction

A repairable system is such that, upon failure it may be put back in the working or operational state by any means other than total replacement of the entire system. This action of making the system operational once more will be called 'repair'. It could involve replacing failed components by working ones, restoring broken connections, mending it or any part of it by machining, cleaning, lubricating, etc. We observe repairable systems all around us. For example, a car upon failure to run, may be restarted by repairing or replacing the failed components such as the battery, break linings, ignition switch, tyres, etc. In the industrial setting, repairing a failed system is a routine activity. In the biological setting systems tend to be one shot affairs, meaning upon failure (death), they remain failed (dead) except in some rare and experimental situations. However, the methodology discussed in this chapter can be used to analyse the occurrence of a sequence of episodes such as epileptic attacks, occurrence of successive tumours and other biological events.

The data from repairable systems consist of the successive failure times of the system: $T_0 = 0 \leq T_1 \leq T_2 \leq T_3 \leq \cdots$. It is assumed that the repairs are instantaneous, and if not, the time taken for repairs is ignored by stopping the clock while the repairs are going on. In case the repair brings the system to the condition of a brand new system, identical in properties to a fresh one, then it would be reasonable to assume that the successive interfailure times,

$$T_1, T_2 - T_1, T_3 - T_2, \cdots, \quad \text{etc.},$$

form a sequence of independent and identically distributed random variables. In such a situation the methods developed earlier (for complete random samples) would also apply here. However, in this chapter we consider other models of repair.

10.2 Repair Models

A repair which changes a failed system into one which is functional, and as good as a brand new system (of age 0) is called a *perfect repair*. This may be considered to be equivalent to replacing a failed system by an identical to a new system.

On the other hand, the repair of the failed system may restore it to its state just prior to failure. That is to say, while the state changes from failed to operational, its properties are those of a system which has the age just prior to its failure. This kind of repair is very common in practice. If the failure of a small but critical component causes the failure of the entire system then the replacement or repair of this component is often the method of repair. The rest of the system continues to be the same old one. Hence it may be realistically assumed that the system has the same age and other characteristics that it had just prior to the failure. This type of repair is termed *minimal repair*. Let us denote by F the c.d.f. of the time to first failure (T_1) of the system. Under the minimal repairs model the conditional distribution of the i-th interfailure time $S_i = T_i - T_{i-1}$ is given by the survival function

$$\overline{F}_t(s) = P[S_i > s | T_{i-1} = t] = \frac{\overline{F}(t+s)}{\overline{F}(t)}.$$

10.3 Probabilistic Models

Let $N(t)$ denote the number of failures upto time t. If we assume the simple conditions for a Poisson process, viz.,

(a) $N(0) = 0$, i.e., there are no failures at time 0,

(b) $[N(a_2) - N(a_1)]$ and $[N(b_2) - N(b_1)]$ for $a_1 < a_2 \leq b_1 < b_2$; which are the number of failures in disjoint intervals $(a_1, a_2]$ and $(b_1, b_2]$ respectively are independently distributed.

(c) There exists a function $\lambda(t)$, called the intensity function such that

$$\lim_{\Delta t \to 0} \frac{P(N(t+\Delta t) - N(t) = 1)]}{\Delta t} = \lambda(t)$$

and

$$\lim_{\Delta t \to 0} P[N(t+\Delta t) - N(t) = 2 \text{ or more}] = 0.$$

The important consequence of the above conditions is that the number of failures in the interval $(t_1, t_2]$, has the Poisson distribution with parameter

$$\int_{t_1}^{t_2} \lambda(u)du = \Lambda(t_2) - \Lambda(t_1), \quad \text{say.}$$

The process $N(t)$ is then said to be a non-homogeneous Poisson process (NHPP) with mean function $\Lambda(t)$, (or intensity function $\lambda(t) = \frac{d}{dt}\Lambda(t)$).

In this formulation, if a failure occurs at time t, and the system is minimally repaired then the process $N(t)$ continues to be an NHPP with the intensity function $\lambda(t)$. If $\lambda(t) = \lambda$, a constant, then $N(t)$ is a (homogeneous) Poisson process and $N(t_2) - N(t_1)$ has the Poisson distribution with mean $(t_2 - t_1)\lambda$.

The intensity function of the process may also be interpreted as the failure rate of the distribution of the time to first failure. Further, under the minimal repair NHPP model, given the time of the occurrence of i-th failure, $T_i = t_i$, the failure rate of the time to next failure, i.e. $T_{i+1} - t_i$, is $\lambda(t), t > t_i$. The power law intensity function given by

$$\lambda(t) = \frac{\beta}{\theta}(\frac{t}{\theta})^{\beta-1}$$

is often used to model the intensity function of minimal repair model.

This model gives an improving situation, i.e., the mean number of failure in $(t, t+\Delta]$ decreases as t increases as long as $\beta < 1$. If $\beta = 1$, the Poisson process models the situation of neither improvement nor deterioration. The deteriorating situation, when the mean number of failures in the interval $(t, t+\Delta]$ increases with t, is modelled by values of $\beta > 1$.

10.4 Joint Distributions of the Failure Times

The observation on the process $N(t)$ may be terminated in two ways.

(a) *Failure truncation*: Stop observing as soon as a predetermined number n of failures has been observed.
(b) *Time truncation*: Stop observing at a predetermined time t_0.

In the failure truncated case the joint p.d.f. of the failure times $T_1 < T_2 < \cdots < T_n$ can be derived in the following manner.

$$f(t_1, t_2, \cdots, t_n) = f_1(t_1).f_2(t_2|t_1) \cdots f_n(t_n|t_1, t_2, \cdots, t_{n-1})$$
$$0 < t_1 < t_2 < \cdots < t_n < \infty$$

where $f_1, f_2, \cdots$ indicate the designated conditional p.d.f.'s. Under the assumptions of minimal repair model, the NHPP with intensity function $\lambda(t)$ leads to

$$\begin{aligned} &f(t_j|t_1, t_2, \cdots, t_{j-1}) = f(t_j|t_{j-1}) \\ &= \lambda(t_j) \exp\{-\int_{t_{j-1}}^{t_j} \lambda(t)dt\}, \quad t_j > t_{j-1}. \end{aligned}$$

The product of these factors leads to

$$f(t_1, t_2, \cdots, t_n) = \prod_{i=1}^{n} \lambda(t_i) \cdot \exp\{-\Lambda(t_n)\}$$
$$0 < t_1 < t_2 < \cdots < t_n < \infty.$$

The distribution of $T_1, T_2, \cdots, T_{n-1}$ given $T_n = t_n$ is particularly useful. The marginal p.d.f. of T_n can be derived as below

$$f_{T_n}(t) = -\frac{d}{dt}P(T_n > t] = -\frac{d}{dt}P(N(t) < n].$$

Now $N(t)$ has Poisson $(\Lambda(t))$ distribution, hence the above is

$$\begin{aligned} &= -\frac{d}{dt}\sum_{j=0}^{n-1} \frac{(\Lambda(t))^j e^{-\Lambda(t)}}{j!} \\ &= \frac{\lambda(t)(\Lambda(t)]^{n-1} e^{-\Lambda(t)}}{(n-1)!} \end{aligned}$$

by differentiation and appropriate cancellation.

Further, conditional on $T_n = t_n$, the failure time of the n-th failure we have

$$f(t_1, t_2, \cdots, t_{n-1}|t_n) = \frac{f(t_1, t_2, \cdots, t_n)}{f_{T_n}(t_n)}$$
$$= \frac{(n-1)! \prod_{i=1}^{n} \lambda(t_i) e^{-\Lambda(t_n)}}{\lambda(t_n) \cdot (\Lambda(t_n)]^{n-1} e^{-\Lambda(t)_n}}$$
$$= (n-1)! \prod_{i=1}^{n-1} \left[\frac{\lambda(t_i)}{\Lambda(t_n)}\right], \quad 0 < t_1 < t_2 < \cdots < t_n.$$

This is the joint p.d.f. of the order statistics of a random sample of size $n-1$ from the distribution with c.d.f.

$$F(t) = \frac{\Lambda(t)}{\Lambda(t_n)}, \quad 0 < t < t_n.$$

In exactly similar manner one can obtain the joint distribution of $T_1 < T_2 < \cdots < T_n$ given that n failures take place in the interval $(0, t_0]$. It is seen to be

$$f(t_1, t_2, \cdots, t_n|t_0, n) = n! \prod_{j=1}^{n} \frac{\lambda(t_j)}{\Lambda(t_0)}, \quad 0 < t_1 < \cdots < t_n$$

which is the joint p.d.f. of the order statistics of size n from the distribution with c.d.f.

$$F(t) = \frac{\Lambda(t)}{\Lambda(t_0)}, \quad 0 < t < t_0.$$

10.5 Estimation of Parameters

(a) *Constant intensity*

If we assume $\lambda(t) = \lambda$, then due to the properties of the (homogeneous) Poisson process, the inter failure times are i.i.d. r.v.'s with the exponential distribution with mean $1/\lambda$. Hence the usual parametric procedures may be adopted for estimating λ or the mean $\mu = 1/\lambda$. The maximum likelihood estimator of μ is $\frac{t_n}{n}$ and a 100 $(1-\alpha)\%$ confidence interval may be based on the chi-square distribution with $2n$ degrees of freedom which T_n has. Similarly, tests of hypotheses regarding λ (or μ) may also be carried out.

(b) *Power Law Intensity* (failure truncation)
Under the intensity function

$$\lambda(t) = \frac{\beta}{\theta}(\frac{t}{\theta})^{\beta-1}, \quad t > 0,$$

the likelihood of the data $t_1 < t_2 < \cdots < t_n$ is

$$f(t_1, t_2, \cdots, t_n) = \frac{\beta^n}{\theta^{n\beta}}(\prod_{i=1}^{n} t_i)^{\beta-1} e^{-(\frac{t_n}{\theta})^\beta}, 0 < t_1 < \cdots < t_n.$$

The log likelihood is

$$L(\theta, \beta) = n \log \beta - n\beta \log \theta + (\beta - 1) \sum_{i=1}^{n} \log t_i - (\frac{t_n}{\theta})^\beta$$

and the likelihood equations are

$$\frac{n}{\beta} - n \log \theta + \sum_{i=1}^{n} \log t_i - (\frac{t_n}{\theta})^\beta \log(\frac{t_n}{\theta}) = 0$$

and

$$-\frac{n\beta}{\theta} + \frac{\beta t_n^\beta}{\theta^{\beta+1}} = 0$$

giving $\theta = \frac{t_n}{n^{1/\beta}}$ from the second. Substituting this in the first leads to

$$\hat{\beta} = \frac{n}{\sum_{i=1}^{n} \log(\frac{t_n}{t_i})}$$

and

$$\hat{\theta} = \frac{t_n}{n^{1/\hat{\beta}}}$$

as the explicit expressions for the maximum likelihood estimators. It is readily seen that $\frac{2n\beta}{\hat{\beta}}$ has the chi-square distribution with $2(n-1)$ degrees of freedom. This result is immediately useful in obtaining confidence intervals for β and testing hypotheses about it. The parameter β is the shape parameter and as already explained indicates whether the failures are becoming more frequent ($\beta > 1$) or less frequent ($\beta < 1$). It is often of interest to test the $H_0 : \beta = 1$, i.e., homogeneous Poisson process, within the family of power law processes.

Using the chi-square distribution with $2(n-1)$ d.f. of the pivotal quantity $2n\beta/\hat{\beta}$ the following two-sided confidence interval is set up:

$$\chi^2_{\alpha/2,2(n-1)}\frac{\hat{\beta}}{2n} < \beta < \chi^2_{1-\alpha/2,2(n-1)}\frac{\hat{\beta}}{2n}$$

where $\chi^2_{\alpha/2,2(n-1)}$ and $\chi^2_{1-\alpha/2,2(n-1)}$ are the lower and upper $100\alpha/2\%$ points of the χ^2 distribution with $2(n-1)$ d.f. If the value 1 falls outside the above interval then we would reject the $H_0 : \beta = 1$. One-sided confidence intervals and tests can be similarly set up.

(c) *Power law intensity (time truncation)*

In this case the likelihood of the data $(N, T_1 < T_2 < \cdots < T_N < t)$, is

$$f(N, t_1, t_2, \cdots, t_N) = \frac{(\frac{t}{\theta})^{N\beta}\exp\{-(\frac{t}{\theta})^{\beta}\}}{N!} \quad N!\prod_{i=1}^{N}\frac{\beta}{t}(\frac{t_i}{t})^{\beta-1}$$

since N, the number of failures upto time t and the failure times $0 < t_1 < t_2 < \cdots < t_N < t$, upto time t are both random, the maximum likelihood estimators of β and θ may then be derived as in the previous section. They are

$$\hat{\beta} = \frac{N}{\sum_{i=1}^{N}\log(t/t_i)} \quad \text{and} \quad \hat{\theta} = \frac{t}{N^{1/\hat{\beta}}}.$$

Here, only conditional (given $N = n$) inference for β and θ is possible. The conditional confidence interval can be set up as

$$\chi^2_{\alpha/2,2n}\frac{\hat{\beta}}{2n} < \beta < \chi^2_{1-\alpha/2,2n}\frac{\hat{\beta}}{2n}$$

and further tests for $H_0 : \beta = \beta_0$ may then be constructed as before. See Rigdon and Basu (2000) for further details.

10.6 Unconditional Tests for the Time Truncated Case

The analysis in section 10.5 above has been carried out under the assumption of failure truncation and, if time truncation is actually carried out then conditional on the knowledge of the number of failures $N = n$ upto the time t of trucnation. This is so because the sampling distributions of the pivotal quantity are available in these cases only. Below we provide unconditional tests (without assuming a fixed number of failures) for the time truncated case based on certain asymptotic results. (Bhattacharjee, Deshpande, Naik-Nimbalkar (2004).)

Let $N(t)$ be the counting process indicating the number of failures upto time t.

As discussed earlier, we assume that it has the structure of a non-homogeneous Poisson process (NHPP) with intensity function $\lambda(t)$. We wish to test the null hypothesis

$$H_0 : \lambda(t) = c.\lambda_0(t), \quad 0 < t < t_0, \quad c > 0$$

where t_0 is the time of truncation of the observation of the failure / repair process, against a joint alternative hypothesis

$$H_1 : \frac{\lambda(t)}{\lambda_0(t)} \quad \text{increases in} \quad (0, t_0).$$

We propose the use of the statistic

$$\begin{aligned} Z^*(t) &= \frac{-1}{\sqrt{N(t_0)}} \left[\int_0^{t_0} \left(\log \frac{\Lambda_0(u)}{\Lambda_0(t_0)} + 1 \right) dN(u) \right] \\ &= \frac{-1}{\sqrt{N(t_0)}} \left[\sum_{i=1}^{N(t_0)} \log \left(\frac{\Lambda_0(t_i)}{\Lambda_0(t_0)} \right) + N(t_0) \right]. \end{aligned}$$

Using the fact that conditionally given $N(t_0) = n$ the distribution of $Z = \sum_{i=1}^{n} \log \frac{\Lambda_0(t_i)}{\Lambda_0(t_0)}$, under the H_0 is chi-square with $2n$ degrees of freedom, we can write the unconditional distribution of $Z^*(t)$ as the following mixture:

$$P[Z^*(t) \leq z] = \sum_{n=1}^{\infty} P[\chi^2_{2n} \leq 2(\sqrt{n}z + n)] \cdot \frac{\{c\Lambda_0(t_0)\}^n e^{-c\Lambda_0(t_0)}}{n!(1 - e^{-c\Lambda_0(t_0)})}, \quad (10.6.1)$$

c being the constant of proportionality and Λ_0 being the integrated failure rate (mean function) under H_0. If the experimenter has the knowledge of expected number of failures upto time t_0, which is $c\Lambda_0(t_0)$, under the H_0, then he can use the lower critical points from this mixture distribution. These critical points z for certain values of $c\Lambda_0(t)$ are tabulated below.

Table 10.1

$c\Lambda_0(t)$	z = critical point closest to size $\alpha = .05$
1	-1.009
1.5	$-\ 1.094$
2	-1.15
2.5	-1.19
3	-1.224
4	-1.278
5	-1.3205
10	-1.4345
15	-1.48
20	-1.505
50	-1.558

These critical points approach the standard normal (lower) 5% critical point, which is -1.65. If the experimenter does not know c and / or $\Lambda_0(t_0)$ then he should use this asymptotic critical point. The statistics Z^* has asymptotically as $t \to \infty$ a standard normal distribution under H_0. The proof is not given here. It depends upon the martingale structure of $M(t) = N(t) - \Lambda(t)$ and the martingale central limit theorem.

Actually as long as $N(t)$ continues to be an NHPP with mean function $\Lambda(t)$, the power of the test is also given by the expression based on the mixture of chi-square probability given in (10.6.1) with $c\Lambda_0(t_0)$ replaced by the $\Lambda(t_0)$ specified by the alternative hypothesis.

It may appear that the convergence to the standard normal critical point is rather slow: it has just approached within 0.1 of the asymptotic value (-1.65) by the time 50 failures are expected. Here it should be kept in mind that the role of $c\Lambda_0(t_0)$, the expected number of failures, is like that of sample size n in standard asymptotics where the limiting distributions are approached as $n \to \infty$. Values of n in the neighbourhood of 50 or 100 are not uncommon for close approximations in that context.

The conventional testing procedure is to condition on the observed number n of failures. Then the exact null conditional distribution of Z, which is χ^2 with $2n$ degrees of freedom, is used to obtain the critical points. See Bain, Engelhardt and Wright (1985). Studies have shown that the unconditional test procedure, using the critical points suggested here, have better power and hence is recommended for use over the conditional procedure.

(Bhattacharjee, Deshpande and Naik-Nimbalkar (2004)).

Illustration 10.1

We consider a subset of the data given in Majumdar (1993) on the failure times of a vertical boring machine. The observations are:

376, 808, 1596, 1700, 1701, 1781, 1976, 2076, 2136, 2172, 2296, 2380, 2655, 2672, 2806, 2816, 2848, 2937, 3158, 3575, 3632, 3686, 3705, 3802, 3811, 4020.

Let the null hypothesis be $H_0 : \Lambda_0(u) = u/1000$.

With truncation time $t = 2000, N(2000) = 7$, the value of the statistic is $Z^* = -1.4152$. The critical point at 0.05 level using Table 1 is -1.15. The null hypothesis is rejected.

R-commands are given in the appendix of this chapter.

For truncation time $t = 4000, N(t) = 26, Z = 28.0509, Z^* = -2.3484$ and the corresponding exact critical points are 36.4371 and -1.278 respectively. The null hypothesis is rejected in all the cases. To guess the nature of cumulative intensity function, we plot the graph of $N(t)$ process. For the truncation time $t = 3000$, the plot showed piecewise cumulative intensity. Figure 10.1 is the plot of intensity process upto truncation time 3000 only.

Let $H_0 : \lambda_0(u) = \frac{u}{1500}, 0 < u < 1500$ and $\lambda_0(u) = \frac{17u}{2000} - 11.75, 1500 < u < 3000$.

We obtain $z^* = 0.2734$. The null hypothesis is accepted using exact and asymptotic critical points.

R-commands are given in the appendix of this chapter.

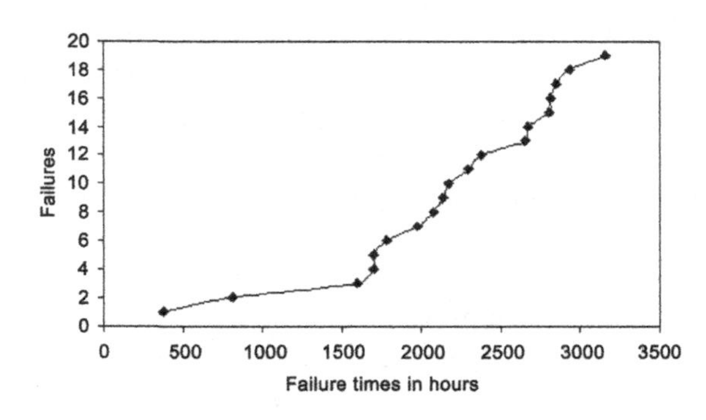

Figure 10.1
Failures of a vertical Boring Machine

Exercise

Exercise 10.1: The following are the failure times of a device in hrs: 80, 120,170, 190, 210, 230, 280, 350, 410, 500, 610, 770, 920, 1090, 1240.
It is known that:

(1) The power low intensity function given by

$$\lambda(t) = \frac{\beta}{\theta}\left(\frac{t}{\theta}\right)^{\beta-1}$$

is appropriate model for the intensity function of the device.

(2) The method of termination of observations used is time truncation with truncation time of 1100 hrs.

Estimate the parameters of the model and construct the conditional (given $N = 15$) confidence interval for the shape parameter β with confidence coefficient of 0.95.

Appendix

R-commands for unconditional test: Boring Machine Failure Data

I. Test of Hypothesis: $\Lambda_0(u) = u/1000$ with truncation time $t = 2000$

R-commands for computation of Z^*

```
> t <-c(376,808,1596,1700,1701,1781,1976);
> # vector of failures;
> t0<-2000;        # Truncation time;
> nt0<-length(t); nt0;
> # Number of failures in the interval (0,2000];
[1] 7
> u<-t/1000;        # Cumulative intensity function;
> u0<-t0/1000;
> # Cumulative intensity function at truncation time;
> s<-sum(log(u/u0)); s;
[1] -3.255795
> m<-(-1/(nt0^.5)); m;
[1] -0.3779645
> z <-m*(s+nt0);     # Computation of test statistic;
> z;      # Value of test statistic;
```

```
[1] -1.415176
```

R-commands for test of hypothesis with truncation time $t = 4000$ are similar. Hence are not given.
Piecewise Intensity Function

```
> t1 <-c(376,808);
> # Vector of failure epochs in the first piece;
> nt1 <-length(t1);
> # Number of failures in the first piece;
> u <-t1/1500;
> to <-3000;      # Truncation time;
> uo <-17 * to/2000 - 11.75;
> # Cumulative intensity at truncation time;
> uo;
[1] 13.75
> s1 <-sum(log(u/uo));  s1;
[1] -7.244367

> t <-c(1596,1700,1701,1781,1976,2076,2136,2172,2296,2380,
  2655,2672,2806,2816,2848,2937);
> # Vector of failure epochs in the second piece.
> nt <-length(t); # Number of failures in the second piece;
> nto <- nt+nt1;  nto;       # Total number of failures;
[1] 18

> u <-17*t/2000-11.75;
> m <- (-1/(nto)^.5);
> m;
[1] -0.2357023

> s2 <-sum(log(u/uo));
> z <-m*(s1+s2+nto); z;    # Value of the test statistics;
[1] 0.2734139
```

Appendix: Statistical Analyses Using R

Many of the statistical methods described in this book can be applied to small data sets with a simple hand calculator. Indeed, it is a good practice to go through these calculations by hand at least once to really understand the way these procedures work. However, when data sets are large and / or the procedures are complex, use of statistical software is a must. Several commercial packages are available. The most widely known are R, SPSS, SAS, S-PLUS, MINITAB etc. We strongly advocate R, a system for statistical analyses and graphics created by Ross Ihaka and Robert Gentleman (1996). R is both a software and a language considered as dialect of language S created by AT&T Bell Laboratories. There are several good reasons for advocating R.

- R is a free software, which makes its use especially in colleges and university courses and also in other settings (as cost factor is important everywhere) very attractive.
- R has an excellent built-in-help system.
- R has good graphing capabilities.
- R is a computer programming language; for new computer users, the next leap to programming is not hard with R and those who are familiar with programming language it will be very easy.
- The language is easy to extend with user-written functions.
- R is compatible with S-PLUS, which is a commercial package. Hence students can easily switch over to it if desired.
- A prominent feature of R is its flexibility. For example, consider regression, a commonly used statistical technique. Suppose one runs a series of 20 regressions and wants to compare the regression coefficients. It is possible to display with R, only the estimated coefficients. Thus the result may take a line or two. However, classical software could well

open 20 results windows.

R installation

R software is obtained from the Comprehensive R Archive Network (CRAN), which may be reached from the R project website at www.r-project.org. The files needed to install R, are distributed from this site where the instructions for installation are also available. There are some packages which are not in the base package of R. Some of these packages you will get along with the base package of R. To see which packages you have, use the command library( ). To install additional packages, under windows environment choose the option "Install packages from CRAN". For this, you must have network connection. If you are working off line, you may use the menu "Install packages from local zip menu file" provided you have already obtained the necessary packages.

How does R work?

R is an object-oriented language. This wording is complex but R language is very simple and flexible. It is interpreted language, meaning that all commands typed on the keyboard are directly executed without requiring building in the complete program like in most computer languages (C, Pascal etc.). Furthermore, R syntax is very simple and intuitive.

Object-oriented means that variables, data, functions, results, etc. are stored in the active memory of the computer in the form of objects with operators (arithmetic, logical and comparison) and functions (which are themselves objects).

Data: Statistics is a study of data. The first thing we should learn is to enter and manipulate the data.

Data-types: The usual data types are available in R and are known as "Modes". The modes are logical (Boolean true/false), numeric (integers and reals), complex (real + imaginary numbers).

Interacting with the interpreter: Data analysis in R proceeds as an interactive dialogue with the interpreter. As soon as we type command at the prompt (>), and press the enter key, the interpreter responds by executing the command.

The R language includes the usual arithmetic operations:

+ : Addition

− : Subtraction

* : Multiplication

/ : Division

^ : Exponentiation

Here are some simple examples of arithmetic in R:

```
> 230 + 540;
[1] 770
```

The symbol [1] in the output indicates a vector. This notation will make sense once vectors are introduced.

```
> 4 ^ 2 - 3 * 2;
[1] 10
> 2 ^ -3;
[1] 0.125
```

It is always better to specify the order of evaluation of the expression by using parenthesis. For example,

```
> 1 - 2 * 3;
[1] -5
> (1 - 2) * 3;
[1] -3
```

Note: Spaces are not required to separate the elements of arithmetic expressions. However, judicious use of spaces can help to clarify the meaning of the expression.

Methods of Data Input:

The Manual Method: R uses the assignment operator <− ("less than" sign followed by "minus" sign) to give a data object (or any other object) its value. The operator −> may also be used. However, with −> operator the assignment is from left to right. As an example consider

```
> x <-2;      # The command assigns the value 2 to object x;
> x^2->y;     # This command assigns the value x^2 to object y;
```

Functions: Many mathematical and statistical functions are available in R. They are also used in the similar manner. A function has a name, which is typed, followed by a pair of parentheses. Arguments are added inside this pair of parentheses as needed.

The mostly useful R command for quickly entering in small data sets is the c ("combine") function. This function combines or concatenates terms together. As an example, consider

```
> y <-c(1, 2, 3, 9, 15, 17); # c function constructs a vector;
```

Note that you must use c function to construct a vector. However, once you have assigned the value to y, you may then reassign its value to other data objects. For example,

```
> z <- y;
```

Acceptable Object Names: We are free to make variable names out of letters, numbers and the dot or underline characters. A name should start with letter and we cannot use any other characters or mathematical operators. Needless to say "case is important."

The c function can also be used to construct a vector of character strings, for example

```
> Names <- c("Ashok","Sandhya","Neela");
```

The data set is stored in R as a vector. This simply means that it keeps track of the order of the components. This is good thing for several reasons:

- It is possible to make changes to the data, item by item instead of having to enter the data set again.
- Vectors are mathematical objects. So the standard arithmetic functions and operators apply to vectors on an element-wise basis. For exmple,

```
> c(1, 2, 3, 4)/2;
[1] 0.5 1.0 1.5 2.0
> c(1, 2, 3, 4) / c(4, 3, 2, 1);
[1] 0.2500000 0.6666667 1.5000000 4.0000000
```

There are other functions, which return vectors as results. For example, the sequence operator (:) generates consecutive numbers, while the seq (sequence) function does the same thing, but more flexibly. Examples are:

```
> 1 : 4;
[1] 1 2 3 4
> 4 : 1;
[1] 4 3 2 1
> -1 : 2
[1] -1  0  1  2
> seq(2, 8, by = 2); # Specify interval and increment;
[1] 2 4 6 8
```

```
> # The sign # specifies a comment;
> # Text to right of # is ignored by interpreter;
> seq(0, 1, length=11);
> # Specifies interval and the number of elements;
[1] 0.0 0.1 0.2 0.3 0.4 0.5 0.6 0.7 0.8 0.9 1.0
```

Warning: Be careful while applying simple arithmetic functions and operators to vectors. If the operands are of different lengths then the shorter of the two is extended by repetition (as in c (1, 2, 3, 4) / 2 above).

If the length of the longer operand is not multiple of the shorter, then a warning message is printed,but the interpreter proceeds with the operation.

```
> c(1, 2, 3, 4) + c(4, 3);
> #c(4,3) is repeated twice(i.e. c(4,3,4,3) is used;
[1] 5 5 7 7
> c (1, 2, 3, 4)+ c(4, 3, 2);
> # c(4, 3, 2) is considered as c(4, 3, 2, 4);
[1] 5 5 5 8
Warning message:
In c(1, 2, 3, 4) + c(4, 3, 2) :
longer object length is not a multiple of shorter object length
```

Modification of the objects: Consider the following simple example

```
> x <-10;
> x;  # Print the value of x;
[1] 10
> x <-2;    x;
[1] 2
```

If the object already exists, its previous value is erased (the modification affects only the objects in the active memory not the data on the disk).

Entering data with scan function (scan( ))

Suppose the body weights (in grams) of 12 rats used in a study of vitamin deficiencies are given:
103, 125, 112, 153, 124, 106, 141, 117, 121, 115, 130, 95.
To create a data object wt using scan function, we type the command,

```
> wt=scan( ); # The following is the response to this command;
1:
```

Now type the data separated by space. When all the 12 values are input, press the enter key twice. Your output will look like:

```
1: 103 125 112 153 124 106 141 117 121 115 130 95
13:
Read 12 items
```

Reading data in a file: For reading and writing in files, R uses the working directory. To know what this directory is, the command getwd( ) (get working directory) can be used.

One of the most straightforward ways to retrieve data is through plain text. Almost all applications used for handling data will export data as a delimited file in ASCII text and this gives us the ready way to get vast majority of data in R.

R can read the data stored in text (ASCII) files with the following functions: read.table, scan( ) and read.fwf.

R can also read other formats (Excel, SAS, SPSS) and access SQL-type databases, but the functions needed for these are not in the package base. These functionalities are useful to more advanced uses of R. However, one can create a file in excel, save it as (delimited) text and read it, in R, using "read.table" command.

Data Frames : A data frame corresponds to what other statistical packages call a "data matrix", or a "data set". It is a list of vectors and / or factors of the same length, which are related "across", such that data in the same position come from the same experimental unit (subject, animal etc). It is possible to create a data frame with the function "data. frame". The vectors so included in the data frame must be of same length, or if one of them is shorter, it is "recycled" a whole number of times.

Squeezing in Big Data Sets: R uses a memory-based model to process data. This means that the amount of data that can be handled is critically dependent upon how much memory is available. Earlier versions required the user to increase the available memory when starting up. But now there is dynamic allocation. However, if you still run out of memory while trying to import a large data set ; you may be able to overcome the problem. Using scan( ) to import file will use less memory. However, scan( ) is not easy to use and you have to enter the column names separately. Going beyond scan( ) there are methods to store your data in a data base table and access the table using the appropriate interface. This enables the user to access huge amounts of data by processing it in bits.

Data Accessing or Indexing: The indexing system is an efficient and flexible way to access selectively the elements of an object; it can either be numeric or logical.

Accessing data from a vector(univariate case): There are several ways to extract data from a vector. Here we present, with a help of an example, a summary,using both slicing and extraction by a logical vector.

Illustration

```
> x <- 2:12; # Assign to vector x the elements 2:12;
> x; # Print the value of the vector x;
 [1]  2  3  4  5  6  7  8  9 10 11 12
> x[3]; # Access ith element of x for i=3;
[1] 4
> x[x > 7]; # All elements >7;
> # This is an example of extraction by logical vector;
[1]  8  9 10 11 12
> x[x > 7 & x <= 11];
> # List of all elements in the interval (7,11];
[1]  8  9 10 11
```

Note: Data accessing is done with square brackects [ ]. It is important to keep this in mind as parentheses ( ) are used for the functions. While extracting specific elements, it is essential to use the c(combine) function. The command x[1,4,7] would mean completely different. It would specify indexing into a three-dimensional array.

Accessing data from a data frame
Illustration

Consider the data on two variables pre and post.
pre : 5260,5470,5640,6180,6390,6515,6805,7515,7515,8230,8770
post : 3910,4220,3885,5160,5645,4680,5265,5971,6790,6900,7335
Input the data.

```
> pre<- c(5260,5470,5640,6180,6390,6515,6805,7515,7515,8230,8770);
> post<-c(3910,4220,3885,5160,5645,4680,5265,5971,6790,6900,7335);
> d<-data.frame(pre,post);
> d;
```

```
    pre post
1  5260 3910
2  5470 4220
3  5640 3885
4  6180 5160
5  6390 5645
6  6515 4680
7  6805 5265
8  7515 5971
9  7515 6790
10 8230 6900
11 8770 7335

> d$pre[3] # Access third element of the variable pre;
[1] 5640
> d$pre # Access the variable pre;
 [1] 5260 5470 5640 6180 6390 6515 6805 7515 7515 8230 8770
```

To access all the elements of the variable pre excepting first, second, third and fourth elements give the command

```
> pre [-c(1,2,3,4)];
 [1] 6390 6515 6805 7515 7515 8230 8770
> d$post[d$post>7315]; #Data extraction by the logical operator(>)
[1] 7335
> d[6,2]; #Access the value in the 6th row and 2nd column;
[1] 4680
> attach(d);
The following objects are masked _by_ .GlobalEnv:
    post, pre
> pre[3]; # 3rd element of the variable pre
[1] 5640
> post [post > 7000]; # Extract all the elements of the;
> # variable post which are larger than 7000;
[1] 7335

Note: The command attach() places the data frame d in the
systems search path. You can view the search path with command;
```

```
> search();
 [1] ".GlobalEnv"          "d"                   "package:stats"
 [4] "package:graphics"    "package:grDevices"   "package:utils"
 [7] "package:datasets"    "package:methods"     "Autoloads"
[10] "package:base"
```

R uses a slightly different method when looking for objects. If the program "knows" that it needs a variable of specific type, it will skip those of other types. This is what saves you from the worst consequences of accidentally naming a variable, say, "c", even when there is a system function of the same name.

Detach: You can remove a data frame from the search path with command detach ( );

```
> detach();
> search();
[1] ".GlobalEnv"          "package:stats"       "package:graphics"
[4] "package:grDevices"   "package:utils"       "package:datasets"
[7] "package:methods"     "Autoloads"           "package:base"
```

Subset and Transform: The indexing techniques for extracting parts of a data frame are logical but a bit cumbersome, and a similar comment applies to the process of adding transformed variables to a data frame. R provides two commands to make things a little easier. The following illustration will explain their use;

```
> data(cars); # Access resident data frame cars;
> cars[1:5,]; # Access first five rows of the data frame cars;
  speed dist
1     4    2
2     4   10
3     7    4
4     7   22
5     8   16
> cars2<-subset(cars,dist>22); # Assign to data object cars2;
```

The subset of the data object cars such that the dist variable is larger than 22. So in the cars2 the first five rows of cars will be definitely removed.

```
> cars2[1:5,]; # Access first five rows of data object cars2;
```

```
   speed dist
8     10   26
9     10   34
11    11   28
14    12   24
15    12   28
> # Observe that rows which dont satisfy the condition;
> # (dist >22) are removed;
> cars3<-transform(cars,lspeed=log(speed));
> cars3[1:5,];
  speed dist   lspeed
1     4    2 1.386294
2     4   10 1.386294
3     7    4 1.945910
4     7   22 1.945910
5     8   16 2.079442
```

Notice that the variables used in the expressions for new variables or for sub-setting are evaluated with variables taken from the data frame. Subset also works on single vector. For example:

```
> data(rivers);
> rivers[1:5];
[1] 735 320 325 392 524
> rivers2 <- subset (rivers, rivers > 735);
> rivers2[1:3];
[1] 1459  870  906
```

Graphics with R: R offers a remarkable variety of graphics. We shall only note here that each graphical function has a large number of options making the production of graphics very flexible and use of drawing package almost unnecessary. The way graphical function works deviates substantially from the scheme sketched earlier. Particularly, the result of graphical function cannot be assigned to a object but it is send to a graphical device. Graphical device is a graphical window or a file.

There are two kinds of graphical functions: the high-level plotting functions, which create a new graph, and low-level plotting functions, which add elements to an already existing graph. The graphs are produced with respect to graphical parameters, which are defined by default and can be modified with the function “par”.

Getting help: The on-line help of R gives very useful information on how to use the function. The help is available directly for a function. For instance:

```
> ?lm ;
```

This command will display, within R,the help for the function lm( ) (linear model). The command help (lm) or help (“lm”) will have the same effect. The last function must be used to access the help with non-conventional characters; for example, the command,

```
> ?* ;    # This command will give the error message. However,;
> help("*"); # Opens the help page for arithmetic operator *.;
```

By default, the function help searches in the packages, which are loaded in memory. The function try.all.packages allows to search in all packages if its value is TRUE. For example,

```
> help("bs");
```

Error in help (“bs”): No documentation for ‘bs’ in specified packages and libraries. You could try ‘help.search(“bs”)’.

```
> help.search("bs");
```

As an output you will see all help files with alias or title matching ‘bs’. (output is not shown)

```
> help ("bs",try.all.packages=TRUE);
```

Topic ‘bs’ is not in any loaded package but can be found in package ‘splines’ in library ‘C:/Program Files/R/R-3.2.2/library’

The function apropos finds all functions whose name contains the characters string given as argument; only the packages loaded in the memory are searched. For example,

```
> apropos("help");
[1] "help"           "help.request" "help.search"  "help.start"
```

The help in html format (read, e.g., with Netscape) is called by typing;

```
> help.start( );
```

A search with keywords is possible with this html help.

Packages: Many R functions are included in the basic system. Other functions must be loaded from libraries. Some of these libraries are distributed with R system. If additional packages are needed, http://www.cran.r-project.org/ is the source for downloads.

Installing and Loading Additional Packages: Follow the following procedure:

1. Keep internet on,
2. In R console, click on packages in the menu bar,
3. From the drop down list, select: Install package(s),
4. From dropdown list select mirror for the session, say India,
5. From the dropdown list select a package to be downloaded, say, mclust,
6. The package and the packages which are dependent on the package of interest will be installed.

To load the installed package, give the command:

```
> library (mclust);
```

Reading the Package Manual and Vignette: After you installed and loaded a new package a good starting point is to read its manual.
For, give the command:

```
> library(help=mclust);
```

Or find the manual on the package website as follows:
Go to R-site, click on packages, choose packages sorted by name, select mclust and download mclust.pdf.
Some package authors also write a vignette, a document that illustrates how to use the package. A vignette typically shows some examples of how to use the functions and how to get started. The key thing is that a vignette illustrates how to use the package with R code and output. To read the vignette for the package, say mclust, try the following:

```
> vignette("mclust");
```

Updating the packages: Most package authors release improvements to their packages from time to time. To ensure you have the latest version use update.package ( ) from package menu on menu bar in R console.

References

Altman, D. G. (1991). *Practical Statistic for Medical Research*, Chapman and Hall, London.

Aras, G. A. and Deshpande, J. V. (1992). Statistical analysis of dependent competing risks, *Statist.*, Dec. 10, 323-336.

Bagai, I., Deshpande, J. V. and Kochar, S. C. (1989). Distribution free tests for stochastic ordering among two independent competing risks, *Biometrika*, **76**, 107-120.

Bain, L.J., Engelhardt, M. and Wright, F.T. (1985). Test for an increasing trend in the intensity of a Poisson process: A power study, *J.A.S.A.*, **80**, 419-422.

Bain, L. J. and Engelhardt, M. E. (1991). *Statistical Analysis of Reliability and Life Testing Models*, Marcel Dekker, New York.

Barlow, R. E. (1968). Likelihood ratio tests for restricted families of probability distributions, *Ann Math Statist.*, **39**, 547-560.

Barlow, R. E. and Campo, R. (1975). Total time on test processes and applications to failure data analysis, Reliability and Fault Tree Analysis, SIAM.

Barlow, R. E. and Proschan, F. (1975). *Statistical Theory of Reliability and Life Testing*, Holt, Rinehart and Winston, INC. New York.

Bartholomew, D. J. (1963). The sampling distribution of an estimate arising on life-testing. *Technometrics*, **5**, 361-374.

Begg, C. B., McGlave, R., Elashoff, R. and Gale, R. P. (1983). A critical comparison of allogenic bone marrow transplantation and conventional chemotherapy as treatment for acute nonlymphocytic leukemia, *J. Clin. Oncol.*, **2**, 369-378.

Bernoulli, D. (1760). Essi d'une nouvelle analyze de la mortalite par la petite verole, et des advantage de l'inoculation pourla preventiv, *Mem. Acad. R. Sci.*.-95.

Bhattacharjee, M., Deshpande, J. V. and Naik-Nimbalkar, U. V. (2004). Unconditional tests of goodness of fit for the intensity of tiem-trucnated non-homogeneous Poisson process, *Technometrics*, **46**, 330-338.

Bickel, P. and Doksum, K. (1969). Tests for monotone failure rate based on normalized spacings, *Ann. Math. Stat.*, **40**, 1216-1235.

Birnbaum, Z. W. and Saunders, S. C. (1958). A statistical model for life-length of materials, *J. Am. Statist. Assoc.*, **53**, 151-160.

Breslow, N. E. and Crowley, J. (1974). A large sample study of the lifetables and product limit estimates under random censorship, *Ann. Statist.* **2**, 437-53.

Box, G.E.P. (1954). Some theorems on quadratic forms applied in the study of analysis of variance problems, I. Effect of inequality of variance on the one-way classification, *Ann. Math. Statist.*, **25**, 290-302.

Burdette, W. J. and Gehan, E. A. (1970). *Planning and Ananlysis of Clinical Studues.* Charles E Thomas, Springfield.

Burns, J. E. et al. (1983). Motion sicknes incidence: distribution of time to first emesis and comparison of some complex motion sickness conditions, *Aviation Space and Environmental Medicine*, 521-527.

Burns, K. C. (1984). Motion sickness incidence, distribution of time to first emesis and comparison of some complex motion condtions, *Aviation, Space and Environmental Medicine*, **56**, 521-527.

Champlin, R., Mitsuyasu, R., Elashoff, R. and Gale R. P. (1983). Recent advances in bone marrow transplantation. In UCLA symposia on Molecular and Cellular Biology. ed. R. P. Gale, **7**, 141-158. Alan R. Liss, New York.

Chaudhuri, G., Deshpande, J. V. and Dharmadhikari, A. D. (1991). Some bounds on reliability of coherent systems of IFRA components, *J. Appl. Prob.*, **28**, 709-714.

Cohen, A. C., Jr. (1959). Simplified estimators for normal distribution when samples are singly censored or truncated, *Techno-*

metrics, **1(3)**, 217-237.

Cohen, A. C., Jr. (1961). Table for maximum likelihood estimates: singly truncated and singly censored sample, *Technometrics*, **3**, 535-541.

Cohen, A. C., Jr. (1963). Progressively censored samples, *Technometrics*, **3**, 535-541.

Cohen, A. C. (1963). Progressively censored samples in life testing, *Technometrics*, **5**, 237-239.

Cohen, A. C., Jr. (1965). Progressively censored samples in three parameter log-normal distribution, *Technometrics*, **18**.

Cohen, A. C. (1976). Progressively censored sampling in three parameter lognarmal distribution, *Technometrics*, **18**, 99-103.

Cox, D. R. (1972). Regression models and life tables. *J. of Royal Statistical Society*, **B34**, 187-220.

Cox, D. R. and Oakes, D. (1984). *Analysis of Survival Data*, Chapman and Hall, New York.

Crowder, M. J. (2001). *Classical Competing Risks*, Chapman and Hall, New York.

Cutler, S. J. and Ederer, F. (1958). Maximum utilization of the life table method in analysing survival, *J. Chron. Dis.*, **8**, 699-712.

Cutler, S. J., Ederer, F., Griswold, M. H. and Greenberg, R. H. (1960). Survival of patients with ovarian cancer, Connecticut 1935-54, *J. Natioanl Cancer Institute*, **24**, 541-549.

David, H. A. and Moeschberger, M. L. (1978). *The Theory of Competing Risks*, Charles Griffin and Company Ltd., London.

Deshpande, J. V. (1983). A class of tests for exponentiality against increasing failure rate average alternatives, *Biometrika*, **70**, 2, 514-518.

Deshpande, J. V., Frey, J. and Oztusk, O. (2005). Inference regarding the constant of proportionality in the Cox hazards model, Proc. Int. Srilankan Statist. Conf.: Visions of Futuristic Methodologies, Eds. B. M. deSilvia and N. Mukhopadhyay, RMIT, Melbourne.

Deshpande, J. V. and Sengupta, D. (1995). Testing the hypothesis of proportional hazards in two populations, *Biometrika*, **82**, 251-261.

Deshpande, J. V., Gore, A. P. and Shanubhogue, A. (1995). *Statistical Analysis of Non-normal Data*, New Age International Publishers Ltd., Wiley Eastern Ltd.

Deshpande, J. V. and Kochar, S. C. (1985). A new class of tests for testing exponential against positive ageing, *J. Indian Statist. Assoc.*, **23**, 89-96.

Dinse, G. E. (1980). Nonparametric estimation for partially complete time and type of failure data, *Biometrika*, **38**, 417-431.

Dixon W. J. and Massey, F. J. (1983). *Introduction to Statistical Analysis,* 4th ed. McGraw Hill, 598.

Doksum, K. and Yandell, B. S. (1984). Tests for exponentiality, *Handbook of Statistics*, **4**, 579-611.

Ebeling, C. E. (1997). *Reliability and Maintainability Engineering*, McGraw Hill, New York, 296.

Efron, B. (1967). The two sample problem with censored data. *Proc. 5th Berkeley Symp.* Vol. 4.

Elandt-Johnson, R. E. and Johnson, N. L. (1980). *Survival models and Data Analysis*, John Wiley and Sons, New York.

Epstein, B. and Sobel, M. (1953). Life testing, *J. Am. Statist. Assoc.*, **48**, 486-502.

Feinleib, M. (1960). A method of analyzing log-normally distributed survival data with incomplete follow-up, *J. Am. Statist. Assoc.*, **55**, 534-545.

Feinleib, M. and MacMohan, B. (1960). Variation in duration of survival of patients with chronic Leukemia, *Blood*, **17**, 332-349.

Fleming, T. R. and Hanrrington, D. P. (1979). *Counting Processes and Survival Analysis*, Wiley, New York.

Fleming, T. R., O'Fallon, J. R., O' Brien, P. C., and Harmington, D. P. (1980). Modified Kolmogorov-Smirnov test procedures with application to arbitrarily right-censored data, *Biometrics*, **36**, 607-625.

Fortier, G. A., Constable, W. C., Meyers, H., and Wanebo, H. J. (1986). Prospective study of rectal cancer. *Arch. Surg.* **121**, 1380-1385.

Freireich, E. J., Gehan, E. A., Frei, E., et al. (1963). The effect of 6-Mercaptopurine on the duration of steroid induced remissions in acute leukemia, A model for evaluating other potentialuseful therapy, *Blood*, **21(6)**, 699-716.

Gajjar, A. V. and Khatri, C. G. (1969). Progressively censored sample from log-normal and logistic distributions, *Technometrics*, **11**, 793-803.

Gehan, E. A. (1965). A generalized Wilcoxon test for comparing arbitrarily singly-censored samples, *Biometrika*, **52**, 203-223.

Gross, A. J. and Clark, V. A. (1975). *Survival Distributions: Reliability Applications in the Biomedical Sciences.* John Wiley and Sons, New York.

Hall, W. J. and Wellner, J. A. (1980). Confidence bands for a survival curve from censored data, *Biometrika*, **67**, 133-143.

Hanagal, D. (2011). *Modelling Survival Data Using Frailty Models*, Chapman and Hall/CRC, Boca Raton.

Hand, D. J., Daly, F., Lunn, A. D., McConway, K. J. and E. Ostrowaski (ed)(1993). *A Handbook of Small Data Sets,* Chapman and Hall, 203.

Harrington, D. P. and Fleming, T. R. (1982). A class of rank test procedures for censored survival data. *Biometrika*, **69**, 553-566.

Harris, C. M. (1968). The Pareto distribution as a queue service discipline, *Operations Research*, **16**, 307-313.

Harris, C. M. and Adelin, A. (1991). *Survivorship Analysis for Clinical Studies*, Marcel Dekker Inc, New York, 37.

Hoeffding, W. (1948). A class of statistics with asymptotically normal distribution, *Ann. Math. Statist.*, **19**, 293-325.

Hoel, D. G. (1972). A representation of mortality data by competing risks, *Biometrics*, **28**, 475-488.

Hollander, M. and Proschan, F. (1972). Testing whether new is better than used, *Ann. Math. Statist.*, **43**, 4, 1136-1146.

Hollander, M. and Proschan, F. (1975). Tests for the mean residual life, *Biometrika*, **62**, 585-594.

Hollander, M. and Proschan, F. (1984). Non-parametric concepts and methods in reliability, *Handbook of Statistics*, **4**, 613-655.

Horner, R. D. (1987). Age at onset of Alzeimer's disease: Clue to the relative importance of ecologic factors?, *Amer. J. Epidemiology*, **126**, 409-414.

Ihaka, R., and Gentleman, R. (1996). R: a language for data analysis and graphics, *J. of Comput. and Graphical Statis.*, **5**, 299-314.

Kamins, M. (1962). Rules for planned replacement of aircraft and missile parts. RANDMemo, Rm - 2810-PR (Abridged).

Kaplan, E. L. and Meier, P. (1958). Non-parametric estimation from incomplete observations, *J. Am. Statist. Assoc.*, **53**, 457-481.

Kiefer, J. and Wolfowitz, J. (1956). Consistency of the maximum likelihood estimator in the presence of infinitely many incidence parameters, *Ann. Math. Statist.*, **27**, 887-906.

Kimber, A. C. (1990). Exploratory data analysis for possibly censored data from skewed distributions, *App. Statist*, **39**, 21-30.

King, M., Bailey, D. M., Gibson, D. G., Pitha, J. V. and MacCay, P. B. (1979). Incidence and growth of mamary tumors induced by a 7-12dimethylbenz(x) antheacene as related to the dietary content of fat and antioxidant, *J. National Cancer Instittue*, **63**, 296-315.

Kirk, A. P., Jain, S., Pocock, S., Thomas, H. C. and Sherlock, S. (1980). Late results of the Royal Free Hospital prospective controlled trial of prednisolone theray in hepatitis surface antigen negative chronic hepatitis, *Gut*, **21**, 78-83.

Klefsjo, B. (1983). Some tests against ageing based on the total time on test transform, *Com. Statist. Theo. Meth.*, **12(8)**, 907-927.

Klein, J. P., and Moeschberger, M. L. (2003). *Survival Analysis: Techniques for Censored and Truncated Data*, Springer Verlag, New York.

Kochar, S. C. (1985). Testing exponentiality against monotone failure rate average, *Com. Statist. Theo. Meth.* **14(2)**, 381-392.

Kochar, S. C. and Carriere, K. C. (2000). Comparing subsurvival functions in the competing risks model, *Life Time Data Analysis,* **6**, 85-97.

Koul, H. L. (1978). Testing for new better than used in expectation, *Com. Statist. A, Theory and Methods*, **7**, 685-701.

Langenberg, P. and Srinivasan, R. (1979). Null distribution of the Hollander - Proschan statistic for decreasing mean residual life, *Biometrika*, **66**, 679-680.

Lemmis, L. M. (1995). *Reliability: Probabilistic Models and Statistical Methods*, Prentice-Hall, New Jersey, 255.

Majumdar, S. K. (1993). An optimal maintenance strategy for a vertical boring machine, *Opsearch*, **30**, 4.

Mantel, N. and Haenszel, W. (1959). Statistical aspects of the analysis of data from retrospective studies of disease, *J. Nat. Cancer Inst.,* **22**, 719-748.

Mantel, N. (1963). Chi-square tests with one degree of freedom extension of the Mantel-Haenszel procedure, *J. Am. Statist. Assoc.*, **58**, 690-700.

Mendenhall, W. and Lehman, E. H. (1960). An approximation to the negative moments of the positive binomial useful in life testing. *Technometrics*, **2**, 227-242.

Millar, R. G. (1981). *Survival Analysis*, McGraw-Hill, New York.

Moore, D. S. (1968). An elementary proof of asymptotic normality of linear functions of order statistics, *Ann. Math. Statist.*, **39**, 1, 263-265.

Nair, V. N. (1984). Confidence bands for survival functions with censored data, a comparitive study, *Technometrics*, **26**, 265-275.

Nelson, J. W. (1969). Hazard plotting for incomplete failure data, *J. Qual. Technol.*, **1**, 27-52.

Osgood, E. W. (1958). Methods for analyzing survival data, Illustrated by Hodgkin's Disease, *Amer. J. Medicine*, **24**, 40-47.

Peterson, A. V. (1977). Expressing the Kaplan-Meier estimate as a function of empirical subsurvival functions, *J. Am. Statist. Assoc.*, **72**, 854-8.

Peto, R. and Peto, J. (1972). Asymptotically efficient rank invariant procedures, *J. Royal Statist. Society*, **A135**, 185-207.

Pocock, S. J., Gore, S. M. and Kerr, G. R. (1982). Long-term survival analysis: the curability of breast cancer, *Statist. Medicine*, **1**, 93-104.

Pocock, S. J. (1983). *Clinical Trials - A Practical Approach*, Wiley, New York.

Puri, M. and Sen, P. K. (1971). *Nonparametric Methods in Multivariate Analysis*, John Wiley and Sons, New York.

Rigdon, S. E. and Basu, A. P. (2000). *Statistical Methods for the Reliability of Repairable Systems*, John Wiley and Sons.

Sengupta, D. and Deshpande, J. V. (1994). Some results on the relative ageing of two life distributions, *J. Appl. Probab.*, **31**, 991-1003.

Shapiro S. S. and Gross A. J. (1981). *Statistical Modelling Techniques*, Marcel Dekker, New York.

Singh, H. and Kochar, S. C. (1986). A test for exponentiality against HNBUE alternatives. *Comm. Statist. Theor. Meth.*, **15(8)**, 2295-2304.

Tarone, R. E. and Ware, J. (1977). On distribution-free tests for equality of survival distributions, *Biometrika*, **64**, 156-160.

Wilcoxon, F. (1945). Individual comparisons by ranking methods, *Biometrics*, **1**, 80-83.

Wilk, M. B., Gnanadesikan, R. and Huyett, M. J. (1962). Estimation of parameters of the gamma distribution using order statistics, *Biometrika*, **49**, 525-545.

Index

Printed in the United States
By Bookmasters